Lecture Notes in Economics and Mathematical Systems 640

Founding Editors:

M. Beckmann
H.P. Künzi

Managing Editors:

Prof. Dr. G. Fandel
Fachbereich Wirtschaftswissenschaften
Fernuniversität Hagen
Feithstr. 140/AVZ II, 58084 Hagen, Germany

Prof. Dr. W. Trockel
Institut für Mathematische Wirtschaftsforschung (IMW)
Universität Bielefeld
Universitätsstr. 25, 33615 Bielefeld, Germany

Editorial Board:

H. Dawid, D. Dimitrov, A. Gerber, C-J. Haake, C. Hofmann, T. Pfeiffer,
R. Slowiński, W.H.M. Zijm

For further volumes:
http://www.springer.com/series/300

Jens O. Brunner

Flexible Shift Planning in the Service Industry

The Case of Physicians in Hospitals

Dr. Jens O. Brunner
Technische Universität München
Lehrstuhl für Technische Dienstleistungen und Operations Management
Arcisstr. 21
80333 München
Germany
jens.brunner@wi.tum.de

ISSN 0075-8442
ISBN 978-3-642-10516-6   e-ISBN 978-3-642-10517-3
DOI 10.1007/978-3-642-10517-3
Springer Heidelberg Dordrecht London New York

Library of Congress Control Number: 2009943836

© Springer-Verlag Berlin Heidelberg 2010
This work is subject to copyright. All rights are reserved, whether the whole or part of the material is concerned, specifically the rights of translation, reprinting, reuse of illustrations, recitation, broadcasting, reproduction on microfilm or in any other way, and storage in data banks. Duplication of this publication or parts thereof is permitted only under the provisions of the German Copyright Law of September 9, 1965, in its current version, and permissions for use must always be obtained from Springer. Violations are liable for prosecution under the German Copyright Law.
The use of general descriptive names, registered names, trademarks, etc. in this publication does not imply, even in the absence of a specific statement, that such names are exempt from the relevant protective laws and regulations and therefore free for general use.

Cover design: SPi Publisher Services

Printed on acid-free paper

Springer is part of Springer Science+Business Media (www.springer.com)

Acknowledgements

I would like to express my special thanks to my advisor Rainer Kolisch who has supported me ever since I started my work under his supervision. In particular, his expertise in the subject of operations management and his modeling skills has helped me to find and formulate an interesting problem for my dissertation. Beside that it is always a great pleasure to work with him. I am obliged to Jonathan Bard who generously agreed to become co-supervisor for my thesis. Especially, I would like to thank him for teaching me many aspects of combinatorial optimization and integer programming techniques. I really enjoyed my time in Austin.

I also would like to take this opportunity to thank my fellow doctoral students and colleagues for the good company and for their help and advice. In this context a special thanks goes to my good friend Christian Heimerl who always offered me his time and understanding and to Günther Edenharter who supported me in my research.

I thank my parents who provided all the opportunities for becoming the person I am at present. Also, I would like to thank my uncle Heinz-Rudi Brunner who is one of the important persons in my life. The same is true for my best friend Jochen Mandl who has continually supported me since our first meeting 5 years ago. Finally, a special thank you goes to my love Ramona Schmitt who always supports and encourages me in my work and especially during the period when I was writing this thesis.

Munich,   Jens Brunner
November 2009

Contents

1 Introduction 1
   1.1 General Economic Situation in Hospitals 1
   1.2 Complexity of Physician Scheduling 1
   1.3 Topic of This Research 2
   1.4 Outline 3

2 Literature Review on Personnel Scheduling 5
   2.1 General Personnel Scheduling 5
   2.2 Physician Scheduling 8
   2.3 Implicit Shift Modeling 9
   2.4 Column Generation and B&P 10

3 MIP Model for Flexible Shift Scheduling of Physicians 13
   3.1 Basic MIP Model 13
      3.1.1 Model Description 13
      3.1.2 Model Formulation 17
   3.2 Model Enhancements 23
      3.2.1 On-Call Services 23
      3.2.2 Time Window Restrictions 27
      3.2.3 Break Assignment 28
      3.2.4 Holidays and Vacations 32
   3.3 Case Study: Anesthetist Scheduling 34
      3.3.1 Current Practice 34
      3.3.2 Solution of the Model 35

4 Solution Methodologies 39
   4.1 Preprocessing 39
   4.2 Heuristic Decomposition Strategy 41
   4.3 Column Generation and B&P Algorithm 44
      4.3.1 Master Problem Formulation 46
      4.3.2 Subproblem Formulation 49
      4.3.3 Finding Integer Solutions 54
      4.3.4 Branching on MP Variables (MPVarB) 55

vii

		4.3.5	Branching on SP Variables (SPVarB)	58
		4.3.6	A Dual Point of View	61
		4.3.7	Heuristics for the B&P Algorithm	63
		4.3.8	Enhancements for the B&P Algorithm	67

5 Experimental Investigations 73
   5.1   Input Data From MRI 73
       5.1.1   Demand Profiles 74
       5.1.2   Basic Parameter Settings 79
   5.2   Heuristic Decomposition 80
       5.2.1   Analysis of Different Model Features 82
       5.2.2   Parametric Analysis 84
       5.2.3   Analysis of Instances of Different Sites 86
   5.3   B&P Algorithm 87
       5.3.1   Two-Week Problems 92
       5.3.2   Four-Week Problems 93
       5.3.3   Six-Week Problems 93
       5.3.4   General Observations 94
   5.4   Comparison of Both Algorithms 94

6 Conclusions and Further Remarks 99
   6.1   Summary and Conclusions 99
   6.2   Final Remarks and Further Research Directions 101

Appendix 103
   A.1   Abbreviations, Notation, and Symbols 103

Bibliography 111

List of Figures

3.1	Definition of variables	17
3.2	Model indices	18
3.3	Definition of shift design constraints	22
3.4	Definition of variables including break assignments	30
3.5	Demand/supply profile corresponding to current practice	35
3.6	Part of the duty roster produced by model for all anesthetists	36
3.7	Demand/supply profile corresponding to model solution	36
4.1	Scheme overview CGA within B&P framework	55
4.2	Master problem (column) variable branching at root node	56
4.3	Subproblem (original) variable branching at root node	60
4.4	Example for initialization heuristic with 2-week planning horizon	65
4.5	Pseudo code rounding heuristic	66
5.1	Average demand per half hour period in 2005	74
5.2	Hospital-wide cumulated demand in 2005	76
5.3	2-week demand in ZOP	76
5.4	Aggregated 2-week demand in ZOP, HNOP, and SPOP (profile 2)	77
5.5	Aggregated 2-week demand in all operating rooms (profile 3)	77
5.6	Demand profile central operating theater first quarter in 2005	78
5.7	Demand profile central operating theater second quarter in 2005	78
5.8	Demand profile central operating theater third quarter in 2005	79
5.9	Demand profile central operating theater fourth quarter in 2005	79
5.10	Progress of LB in the root node for problem 8	97
5.11	Progress of UB in the course of the algorithm for problem 8	97
A.1	Holidays in 2005	108
A.2	On-call hours counted to regular working hours for each on-call service in 2005	108
A.3	Duty roster produced by model for all anesthetists	109

List of Tables

3.1	Undertime results for 2-week example	36
3.2	On-call assignments for 2-week example	37
4.1	First week solution	43
4.2	Second week solution alternative 1	44
4.3	Second week solution alternative 2	45
4.4	Example REC	57
4.5	Example branching strategy SPVarB	59
4.6	1-pattern	59
4.7	2-pattern	59
4.8	3-pattern	59
4.9	Example of the rounding heuristic	67
5.1	Average day demand on Monday through Friday in 2005	75
5.2	Basic parameter settings	80
5.3	Problem size for weekly model	81
5.4	Comparison of model features	81
5.5	Computational results for parameter $\overline{T}_j^{shift}$	82
5.6	Computational results for parameter T_j^{win}	82
5.7	Computational results for number of physicians	83
5.8	Problem size for weekly model for different demand profiles	83
5.9	Computational results for different demand profiles	83
5.10	Parameter settings for arbitrary FT	87
5.11	Parameter settings for arbitrary PT	88
5.12	Results for 2-week problems	89
5.13	Results for 4-week problems	90
5.14	Results for 6-week problems	91
5.15	Computational results different parameter settings	95

Chapter 1
Introduction

The introductory chapter consists of four sections. In Sect. 1.1 we reveal the current situation in hospitals that is faced by the management. We address the general issue of personnel scheduling in the service industry in Sect. 1.2. Then we motivate our research by considering physicians as the scheduling object. In particular, we show the complex nature of physician scheduling in a hospital environment. The focus of the research is presented in Sect. 1.3. Finally, we conclude the chapter by illustrating the outline of the thesis.

1.1 General Economic Situation in Hospitals

The mounting pressure in the health care industry to reduce costs is forcing hospitals and related facilities to take a closer look at their staffing policies (see [111]). A primary difficulty in reducing personnel costs, the major component of the budget, is the variability in demand and the need to assign staff to fixed shifts. Furthermore, government run facilities, especially those in the European Union, are seeing their budgets cut in terms of real dollars despite an aging and more acutely ill patient population (e.g., see [96]). It has been reported that up to a third of the hospitals in Germany plan a reduction in staff (see [91]). The scheduling process is further complicated by the generally recognized importance of taking individual preferences into account. More attractive schedules promote job satisfaction, increase productivity, and reduce turnover (cf. [2]). However, without improved scheduling procedures that better match supply to demand, the level of care that they now provide will soon become unsustainable.

1.2 Complexity of Physician Scheduling

What makes personnel scheduling in the service sector so much more difficult than in other areas is the wide fluctuations in demand that typically occur throughout the day, and from one day to the next (e.g., see [64]). In health care, much of the

J.O. Brunner, *Flexible Shift Planning in the Service Industry*, Lecture Notes
in Economics and Mathematical Systems 640, DOI 10.1007/978-3-642-10517-3_1,
© Springer-Verlag Berlin Heidelberg 2010

related research has concentrated on nurse scheduling with an emphasis on developing monthly rosters. Effective methods now exist that address both the midterm and short-term problems, and have been implemented in many facilities with some degree of success (e.g., see [17, 31, 41, 49, 82]). The problem of physician scheduling has received much less attention, and for the most part, is still done manually at great time and expense. The complex nature of physician scheduling makes it difficult to define a set of generally accepted hard or even soft constraints, as is the case with nurse scheduling. Accommodating individual nurse preference has long been a consideration of unit managers but little work has been done with respect to physicians. The challenge of dealing with physicians stems from their specialized skills and the near-monopolist environment in which medical services are provided. This gives them disproportionate leverage to negotiate favorable employment terms that are not always in the best interest of the health care organization and the general public. Hence, hospitals often have detailed labor contracts with their physicians that vary by region, governing authority, seniority, specialty, and training that discourage generalization. To date, it has not been possible to successfully define a generic problem that lends itself to standard modeling approaches that rely on a set of basic shifts and start times (e.g., see [14, 23]).

When constructing midterm schedules, it is necessary to plan for regular coverage over the week as well as for on-call services to handle emergencies in off hours. The flexibility that exists in many general labor contracts to start shifts of different lengths at various times, increases the options for constructing lines-of-work. In addition, as mentioned, individual agreements between physicians and the hospital further compound the scheduling process by raising issues of equity and fairness. Considering all scheduling characteristics by staff planners is virtually impossible but might lead to big cost savings for the health care organization. The main issue is the size of the solution space and consequently there is need for more efficient computational methods.

1.3 Topic of This Research

As a consequence of the complex nature of the physician scheduling problem, we have developed a new approach that treats the construction of shifts implicitly by building them period by period over the day. We call this *flexible shift scheduling* and demonstrate its effectiveness for scheduling physicians in the anesthesia department of a university hospital in Munich, Germany – Klinikum Rechts der Isar (MRI). Since the beginning of 2007, German university hospitals have been subject to a newly enacted labor contract that regulates and reforms the common working conditions for physicians (see [88]). In part, the contract defines how to construct feasible lines-of-work that take into account regular hours, overtime, and individual arrangements between the health care facility and the physicians. Rules are also included for scheduling on-call (stand-by) services, which are above and beyond the normal shift. As mentioned, these services are necessary to handle off-hour emergencies and unexpected call outs.

The current system at MRI has grown haphazardly over the years and has become self-perpetuating. In most hospitals, there is a strong hierarchy that makes it difficult to implement new procedures because someone in authority is always disadvantaged in the process. The interest at MRI in a more automated approach arose from the need to abide by the new labor agreements, which are violated by the schedules being produced today. The violation of labor agreements is a general problem that is faced by hospital managers (see [75]). As the new requirements were phased in, mangers were finding it increasingly difficult to satisfy them. As a consequence a generic system is required that can be easily adjusted to new emerging influences in the future. In response, we were asked to develop an efficient scheduling algorithm that addressed not only the hard constraints in the labor contract, but the softer "people" issues as well. To some extent, our model can accommodate a range of individual preferences that serve to improve morale and increase job satisfaction.

The objective of the problem addressed is to find high quality rosters that minimize the total assignment cost required to cover forecast demand. The cost consists of paid out time, planned overtime, and the cost for outside resources to cover shortfalls in coverage for regular shift as well as for off-hour periods that are covered with on-call services. As is typically the case, regular time can be viewed as a fixed cost determined by individual contracts. In German university hospitals, each physician must be assigned anywhere from 42 to 62 h a week depending on his contract, but is rostered for up to several weeks in advance. Our focus is on midterm planning where biweekly or monthly schedules must be generated that meet both system-wide and personal constraints. The problem is modeled as a *mixed integer program* (MIP) in which the construction of shifts is treated implicitly. Rather than starting with predefined weekly templates or shift types, the solution process itself generates shifts and lines-of-work for each physician. Introducing more flexibility in personnel scheduling, can lead to significant reductions in staffing costs (e.g., see [13] or [106]). The approach to the implicit modeling of shifts is a novel way of accommodating various types of flexibility and preference considerations in a problem, and carries over to the service industry as a whole. The usefulness of the approach can be evaluated by the quality of the results, especially when compared with current practice. To solve the MIP model we have developed a heuristic decomposition that decomposes the problem by week and a *branch-and-price* (B&P) approach (e.g., see [20, 117]) capable of solving large real-life problems to optimality. This involves the decomposition of the original MIP into a set covering-type master problem and a series of physician-based subproblems to generate promising schedules in the form of columns.

1.4 Outline

The thesis is structured in six chapters as follows. In Chap. 2 we review the relevant literature for our research and characterize our problem in the context of personnel scheduling. The chapter consists of a discussion of general shift scheduling literature, a focus on physician scheduling, a review of implicit shift modeling approaches, and an overview of relevant B&P work with an emphasis of solving large-scale personnel scheduling problems. In Chap. 3 we start by summarizing the problem characteristics. Afterwards, we present the basic modeling idea to define shifts that allow for maximum flexibility (see Sect. 3.1) and state the problem as a MIP. In the following Sect. 3.2 we show several model enhancements like the incorporation of on-call services, implicit lunch-break assignments, and starting time windows for physicians to tackle realistic problems. In Sect. 3.3 we present experimental investigations using a real demand profile for one of the operating theaters at MRI. Final rosters are for 2-week planning horizons. Additionally, we show the high quality of the output by comparing it with current practice.

In Chap. 4 we focus on solution methodologies. We start with some preprocessing issues that take advantage of individual problem characteristics to reduce overall problem size. In Sect. 4.2 we present our effective heuristic decomposition scheme that divides the overall MIP into weekly or any multiple of weekly subproblems. Eventually, we conclude the chapter with the presentation of our B&P algorithm. First, we show the basic column generation approach to solve the linear relaxation at the root and successor nodes in the search tree (see Sect. 4.3). Then, in Sects. 4.3.4 and 4.3.5 we show two embodied branching strategies and how to modify the column generation approach to remain valid. Finishing Sect. 4.3 we discuss a starting heuristic for a warm-start of the algorithm and thematize some extensions to improve performance of the solution method.

In Chap. 5 we start by presenting the data used in our computations in Sect. 5.1. In the following we show the results of three computational studies. First, we investigate the proposed heuristic decomposition approach and compare it to solving the overall MIP with the standard optimization program CPLEX. Computational issues are investigated for different scenarios, a range of parameter settings, and the problem size measured in the number of operating rooms. Second, in Sect. 5.3 we focus on solving realistic problem instances with our B&P algorithm. Outputs for up to six-week schedules are reported. Finally, in Sect. 5.4 we solve some of the studied instances in Sect. 5.2 once more but this time with our B&P algorithm. So, the focus is to compare the heuristic decomposition strategy with the B&P approach. We close the thesis in Chap. 6 with a summary of the results and some suggestions for further research directions.

Chapter 2
Literature Review on Personnel Scheduling

In this chapter we review the relevant literature for our work. In Sect. 2.1 we start with reviewing the general personnel scheduling literature and some classification schemes. In Sect. 2.2 we summarize the relevant work on physician scheduling. Afterwards we review in Sect. 2.3 work on implicit modeling techniques which is required for the new implicit shift modeling approach to schedule physicians in a hospital given in Chap. 3. Finally, we conclude in Sect. 2.4 with a focus on research that considers column generation approaches and B&P methodologies for personnel scheduling problems in the service industry. This is the relevant literature for Chap. 4 where we present a new column generation and B&P approach for the flexible shift scheduling of physicians in a hospital.

2.1 General Personnel Scheduling

There is a vast amount of literature on personnel scheduling so our review will be limited to the most relevant work. An introductory tutorial to staff scheduling is given in Blöchliger [34] that presents the basic concepts of the scheduling problem and discusses some facets of staff scheduling. To avoid any ambiguity, we introduce the following three definitions. For the first and the second definition we give an example that shows a binary representation of the respective verbal definition.

- *Shift*: A set of consecutive periods within a day. Its length is the total amount of time it covers (see [14]).
  Example: 1-day planning horizon with 24 periods/hours
  0 0 0 0 0 0 0 1 1 1 1 1 1 1 1 0 0 0 0 0 0 0 0 0
- *Roster*: A combination of shifts and days off assignments that covers a fixed period of time; a line-of-work (see [99]).
  Example: 3-day planning horizon with 72 periods/hours
  0 0 0 0 0 0 0 1 1 1 1 1 1 1 1 0 0 0 0 0 0 0 0 0 (day 1 shift)
  0 0 0 0 0 0 0 0 0 0 0 0 0 0 0 0 0 0 0 0 0 0 0 0 (day 2 off)
  0 0 0 0 0 0 0 1 1 1 1 1 1 1 1 0 0 0 0 0 0 0 0 0 (day 3 shift)

J.O. Brunner, *Flexible Shift Planning in the Service Industry*, Lecture Notes
in Economics and Mathematical Systems 640, DOI 10.1007/978-3-642-10517-3_2,
© Springer-Verlag Berlin Heidelberg 2010

- *Flexibility*: For each employee, the ability to assign arbitrary shift lengths, shift starting times, and break periods, and to accommodate individual preferences, requests, and constraints.

Note, that in the second example we would get one binary sequence of length 72, however we used three lines to facilitate presentation. In the service industry, some of the most studied applications of personnel scheduling include call center staffing, airline crew pairing, nurse rostering, and postal worker tour construction (e.g., see [5, 14, 41, 45, 49, 54, 65, 83, 87, 107, 118]). Ernst et al. [65] classify the related problems into six planning modules that address demand forecasting, days off scheduling, shift scheduling, line-of-work construction, task assignment, and staff assignment. The first module, demand forecasting, determines the requirements per period or per shift. It distinguishes between a task-based view and a flexible view. Days-off scheduling (second module) deals with how rest days are interspersed between work days. Shift scheduling selects from a large pool of candidate shifts to cover the demand over some planning horizon. Candidates are constructed in association with general and individual work regulations. The line-of-work construction module combines days off scheduling with shift scheduling and uses the results to build feasible work patterns (tours) over the planning horizon. The task assignment module then assigns tasks to shifts. The last module, staff assignment, involves the assignment of staff to lines-of-work. Normally, these modules are considered sequentially by planners. In an annotated bibliography Ernst et al. [64] focus mainly on algorithms for personnel scheduling. A collection of some 700 references is classified according to the type of problem type, application area, and method. An earlier classification scheme of general manpower problems is introduced by Tien and Kamiyama [112]. The review classifies the formulations of the problems into five stages, namely, the determination of temporal manpower requirements, total manpower requirement, recreation blocks, recreation/work schedule, and shift schedule.

With respect to the classification scheme by Ernst et al. [65], the majority of the research on personnel scheduling has focused on modules two and three, shift scheduling and days off scheduling, with the objective of minimizing total cost. In what follows we will review the most relevant literature in this area.

Alfares [4] provides an efficient algorithm for the rostering or tour scheduling problem that assigns shifts and two consecutive days off to employees whereas Baily [8] integrates the shift and days off scheduling in a *linear program* (LP). The main model objective is to minimize the cost of premium pay. Compared to a heuristic approach savings of almost 14% are reported. Following the approach of Burns and Carter [43] one has first to develop lower bounds on the workforce size and then introduce them as an additional constraint in a linear programming model. This is sufficient to ensure integer solutions. Research that considers especially the staffing decision which is not the focus of this research is given in Baker Magazine [12], Burns and Narasimhan [42], and Mundschenk and Drexl [94].

Beaumont [23] takes an expansive view of shift scheduling and includes worker availability, the maximum number of workers who can start at the same time, the relative efficiency of a worker, the cost of making a customer wait, annual leave

2.1 General Personnel Scheduling

factors, the expected number of jobs an employee can complete in each period, the number of contractors, and the maximum number of jobs that can be done in each period, as parameters in his model. He also permits a limited amount of queuing of customers, but ultimately, real-world problems turn out to be too large to solve. Similar problems, for instance, are studied by Azmat et al. [7] and Ingolfsson et al. [79]. In another research Bailyn et al. [10] consider self-scheduling as an option. The study finds out that self-scheduling might have positive effects for both nurses and nurse managers. But everyone loses when the objective is not to balance between individual and unit benefits.

When demand is constant over some period cyclic scheduling might be the right choice. Baker [11] presents a simple staffing model that assigns days off to full-time staff given a cyclic seven-day demand pattern. A vast amount on literature that tackles the cyclic scheduling problem can be found, among others, in Bard and Purnomo [18], Bartholdi et al. [21], Beliën and Demeulemeester [29], Hochbaum and Levin [77], Millar and Kiragu [92], and Purnomo and Bard [99].

Looking at advanced computational approaches, Brusco [37] evaluates the performance of dual all-integer cutting planes for solving the tour scheduling problem. He shows that a cutting plane enhanced by a linear programming objective cut and a sophisticated row selection rule improves solution times when compared to a commercial *branch-and-bound* (B&B) code. Bard and Purnomo [16] use a column generation algorithm to investigate the problem of individual preference considerations in the midterm scheduling process. A similar problem is investigated by Azaiez and Al Sharif [6]. A 0-1 goal programming model is solved for a 6-month period. The model takes into account nurse preferences like ratio of night shifts and weekends off as well as hospital objectives as ensuring a continuous service with appropriate nursing skills. Jaumard et al. [81] use an exact branch-and-price algorithm to solve a similar problem and were successful for small size problems. Dowling et al. [59] develop a software package based on simulated annealing to schedule ground staff at a large international airport. They start with a feasible solution and try to improve on it until some termination criteria is fulfilled. Tabu search has also been used frequently for midterm personnel scheduling problems, especially for nurses (e.g., see [32, 39, 40, 60]). Other computational methods as various heuristics, stochastic programming or decomposition approaches have been in the focus of the research community too (e.g., see [15, 27, 36, 44, 50, 68]).

Planning for overtime, our chief concern, dates back to at least the work of Mc Manus [89] who investigates how best to allocate overtime in the British Post Office. His main goal was to identify the optimal level of staffing for given fluctuations in daily demand. By making assumptions about how the workload changes from day to day, he is able to derive a series of optimal scheduling rules. Easton and Rossin [63] argue that the increasing per capita labor expenses have forced service sector employers to increase the use of overtime and decrease the use of part-time labor. They evaluate the effects of alternative overtime staffing and scheduling policies on critical performance measures, such as total labor costs, labor utilization, and workforce size and find out that using overtime with premium pay provides significant savings. An important conclusion was that the ideal workforce size and proportion of overtime allocated for a given scheduling policy seems to be relatively insensitive to changes in per capita labor costs.

2.2 Physician Scheduling

While much has been written on nurse scheduling, only a handful of published research exists on the more complex problem of physician scheduling. There are several reasons for this imbalance. To begin, the demand for nurses is relatively constant over a shift, but for physicians it can fluctuate widely. This makes it more difficult to match supply with demand. In addition, more complex labor agreements and individual contract clauses that physicians are able to negotiate make their scheduling problem less general. Thirdly, shifts for physicians can start at different times depending to some extent on operating room bookings, but nurses work standard 8-h or 12-h shifts with little or no variability in their starting times. Fourthly, the need to account for on-call service requirements adds a complicated series of constraints to the physician scheduling problem. For the most part, nurse scheduling is uniform from one hospital to the next, whereas physician scheduling is much more hospital-centric.

Beaulieu et al. [22] are one of the first to develop a mathematical model for scheduling emergency room physicians. They formulate the problem as a mixed integer program with three distinct 8-h shifts and allow requests for vacations, days off, and particular shifts. Attempts to solve the model with a commercial code for a 6-month planning horizon are not successful so they divide the planning horizon into six 4-week periods. Their decomposition strategy is employed to generate schedules for up to 20 physicians including five with part-time contracts. In related studies, Franz and Miller [70] develop a MIP model for scheduling residents at a large teaching hospital for training purposes. They use a rounding heuristic to get good schedules and provide an adjustment procedure to resolve infeasibilities inherent in the problem, while Cohn et al. [52] present results derived from mathematical programming techniques used to schedule the teaching phase of special training programs for medical residents at Boston university school of medicine. Their method provides one-year schedules and includes the consideration of on-call service as well as a possibility to accommodate vacation requests by each resident. White and White [119] address the problem of scheduling hospital rounds by speciality for teams comprising senior, junior, and resident physicians as well as interns. Tabu search coupled with a logic constraint formalism is used to provide monthly schedules. Ovchinnikov and Milner [97] describe an implementation of a spreadsheet model for assigning medical residents over a 1-year horizon in radiology at the university of Vermont's college of medicine. They argue that spreadsheets are preferable to free standing codes for small size problems, especially when practitioners are the ultimate user.

Sherali et al. [106] investigate the problem of allocating night shifts to residents taking into consideration departmental staffing and skill requirements as well as resident preferences. The underlying MIP is too difficult to solve exactly so several heuristics are developed. Topaloglu [113] use goal programming to model a monthly shift scheduling problem for emergency medical residents. His formulation is limited to two fixed shift types for demand – a 10-h day shift and a 14-h night shift, but allows for scheduling a teaching phase. The model is coded in OPL Studio 3.7 and the largest instances solved contains 5,633 constraints and 5,068 integer variables, which reflects the problem size for 23 residents and a 31-day planning horizon. In a subsequent work more insights to the problem and a real implementation at a pulmonary unit of a local hospital are discussed (see [114]). Rousseau et al. [101] tackle the physician scheduling problem with a combination of three approaches that includes constraint programming, local search, and genetic algorithms. Carter and Lapierre [46] analyze current procedures for scheduling emergency room physicians at six Montréal hospitals and provide a collection of rules and recommendations to improve them.

2.3 Implicit Shift Modeling

An observation from the above given literature is that virtually all models are based on an explicit representation of a shift. In particular, pre-defined shifts are used as an input to the problem and then used as the basic element of constraint construction. Three 8-h shifts, two 12-h shifts and various combinations are commonly used in practice to cover the demand during an arbitrary day. A few authors, though, have tackled the shift scheduling problem, at least in part, with implicit modeling techniques. Furthermore, Bechtold and Jacobs [26] show that implicit optimal modeling for shift scheduling has size and runtime benefits over the general applied set covering method for shift scheduling. For a single-day problem, Thompson [110] developed a linear model built on the premise that if a shift starts in a particular period then it must end in one of several successor periods. In this approach, solutions provide starting and ending periods for the minimum number of required shifts but not the shifts themselves. In a post-processing step, he applies a first-in-first-out rule to construct the actual shifts. The approach derives from the work of Moondra [93] who defined variable sets for the number of shifts starting and ending in each period and formulated constraints to impose limits on their allowable durations. Bechtold and Jacobs [24] extended this idea in the development of an implicit modeling scheme for including lunch breaks in shifts. In a subsequent work Bechtold and Jacobs [25] investigate shift length and break placement flexibility on labor utilization. Addou and Soumis [1] generalize the ideas intorduced by Bechtold and Jacobs [24] and confute the hypothesis that the implicit modeling is only valid when no extraordinary overlap exists. To generate weekly rosters with two rest days, Cezik et al. [47] combined Thompson's modeling technique with days off scheduling requirements in a network flow framework. Outputs were weekly rosters with two days off. Burke et al. [40] worked with time interval-based demand and allowed for the possibility of splitting and then combining different shifts to generate new schedules within a tabu search framework. Greater flexibility was achieved as the time interval was reduced.

Bailey and Field [9] present a linear program that uses 6-, 8-, and 10-h shifts rather then a standard 8-h shift to cover demand periods. Tests of the Flexshift model on a data set of 12-h days report an average gain in labor of 24.2%. Some other research studies consider also implicit modeling techniques (e.g., see [38, 80, 100]).

2.4 Column Generation and B&P

The success story of *column generation* (CG) to solve large-scale problems dates back to middle of the 20th century. General reviews that discuss column generation are given in, for example, Wilhelm [120], Soumis [109], and Desrosiers [58]. The idea of implicit dealing with the decision variables in a multicommodity flow problem is introduced by the work of Ford and Fulkerson in the 1950s (see [69]). Dantzig and Wolfe [53] use this idea to extend linear programs columnwise on an as needed basis in the solution process. Then, the cutting stock problem is the first real application where this fundamental work has been implemented (see [73, 74]).

Savelsbergh [104] concentrates on solving the general assignment problem using column generation embedded in a B&B framework to achieve integer solutions. Compared with this Vanderbeck and Wolsey [117] present a general framework that can simultaneously deal with columns and right-hand side integer vectors. Additional they give advice to use lower bounds and to reduce the tailing off effect. Other applications dealing with column generation solving *integer programs* (IPs) are described in Al-Yakoob and Sherali [3], Bard and Purnomo [16], Sarin and Aggarwal [103], Purnomo and Bard [99], as well as in Gamache et al. [71].

Desaulniers et al. [56] present various accelerating strategies to speed up a column generation algorithm. However, the application is restricted to a class of vehicle routing and crew scheduling problems. Desaulniers et al. [56] concentrate on each step of the solution process, namely preprocessing, subproblem, master problem, branch-and-bound and postprocessing. Nevertheless, the strategies and suggestions can easily be adapted to other research areas like physician scheduling, the focus of our work.

Barnhart et al. [20] review the B&P literature and attempt to derive a general methodology. They present several MIPs whose structure is amenable to B&P and discuss major difficulties with respect to the modeling and branching. They also provide valuable insights on implementation and computational issues. Jaumard et al. [81] are the first to propose an exact B&P algorithm to solve the midterm scheduling problem for nurses and where successful for units with up to a dozen nurses. Their main contribution is in the formulation of the subproblem as a resource constrained shortest path problem and in the efficiency of the solution methodology. Rather than modeling resources on the vertices as in many routing applications they model them on the arcs. Taking the same approach, Beliën and Demeulemeester [30] present an integrated operating room and nurse scheduling procedure based on

2.4 Column Generation and B&P

column generation. In the decomposition, one subproblem generates new roster for the nurses while a second produces new workload patterns for the surgery unit. The former is formulated as a shortest path problem with side constraints and is solved with dynamic programming techniques. It is called more often than the latter, which has no special structure, and hence the solutions to the IP formulation are obtained with standard software (CPLEX). Integrality is only forced on the workload patterns when creating the search tree. To find feasible solutions, the master problem is solved as a MIP at each node of the tree after no more column(s) price out favorably. Final results provide lower and upper bounds on the number of nurses required to staff the operating theaters. A broad study is carried out using data from the authors' earlier work (see [28]).

Bard and Purnomo [16] use a column generation scheme to solve the midterm scheduling problem for nurses with an emphasis on preference considerations. Their set covering formulation includes five different shift types and employs a double swapping heuristic to generate promising columns. The cost of a new column is calculated based on the degree to which a roster violates individual preferences. Computational results using data provided by a US hospital show that high-quality rosters can be obtained within a few minutes. In subsequent work, Bard and Purnomo [17] extend their approach to allow for downgrading in the shift scheduling process.

In a study using techniques similar to those presented in Sect. 4.3, Purnomo and Bard [99] propose a new model for the cyclic preference scheduling problem and are able to derive rosters for up to 200 nurses at a time. Their B&P algorithm exploits different branching strategies to limit the size of the search tree and applies an extremely effective rounding heuristic to find integer solutions. The efficiency of the proposed approach is enhanced by applying double aggregation to the rotational profiles that define the subproblems. Runtimes of no more than 10 min are reported for almost all instances investigated. The problem associated with the daily adjustments of midterm schedules is discussed in Bard and Purnomo [19]. A 24-h planning horizon is considered and is solved in a rolling way each 8 h but just the first 8 h are implemented. To solve the problem a B&P algorithm is presented that takes advantage of two branching strategies. The first tackles fractional assignment variables in master problem but is implemented on subproblem variables. The second is associated with slack variables in master problem. The two basic components of the algorithm that contribute most to its effectiveness are a feasibility heuristic to find high-quality integer solutions quickly and a mixed-integer rounding cut procedure to improve on the lower bounds in the search tree. Results are shown for up to 200 nurses. Various other interesting studies that deal with column generation are given in Ceselli and Righini [48], Dumas et al. [61], Easton and Rossin [62], Eveborn and Ronnqvist [66], Hoffmann and Padberg [78], Lavoie et al. [85], Lübbecke and Desrosiers [86], Mehrotra et al. [90], and Ni and Abeledo [95].

Our research falls in the combined categories of shift scheduling, days off scheduling, and line-of-work construction, as described by Ernst et al. [64]. We take an integrated approach that allows for flexible start times, variable shift lengths, break inclusion, overtime and on-call service, shift spillover from one day to the next, and various individual preferences and constraints. One of the primary contributions of the proposed implicit modeling approach is that it allows for flexible starting times and variable shift lengths. In addition, it treats breaks implicitly rather then explicitly, which is a more efficient way of including them in the formulation. In the second part of the research we focus on solution methodologies. First, an efficient decomposition heuristic is presented and then we show a B&P algorithm to solve the underlying MIP. Several branching strategies and methods to find high quality integer solutions are reported to reduce the size of the search tree and to terminate with an optimal solution. To the best knowledge there is no other work in the personnel scheduling literature that combines implicit shift modeling ideas with B&P techniques to solve shift scheduling problems, especially when physicians are the main scheduling object.

Chapter 3
MIP Model for Flexible Shift Scheduling of Physicians

In this chapter we present the general problem of physician scheduling. As will be seen the modeling ideas can be carried over to the service industry as a whole. In Sect. 3.1 we start by stating the problem characteristics and then formulating the basic problem as a MIP where we introduce our new implicit modeling approach. Afterwards in Sect. 3.2 we discuss some extensions to the basic model to tackle real world problems. Eventually, in Sect. 3.3 we close the chapter by presenting an example of the final rosters and compare the output to current practice to show the effectiveness of the proposed approach.

3.1 Basic MIP Model

In the next two sections we describe the general problem structure and then derive a mathematical formulation of the basic problem.

3.1.1 Model Description

Our model for physician scheduling has been developed in cooperation with MRI which has different groups of physicians like full- and part-time physicians. To facilitate the presentation of the problem we take the simplest version of the problem where the workforce being scheduled is assumed to consist of regular full-time employees with identical qualifications. We begin by identifying the guidelines and constraints for this version and then discuss extensions to cover more complex scenarios in Sect. 3.2. Additionally, to reduce the use of notation, some data values such as minimum and maximum shift lengths will be stated explicitly rather than as parameters, but it should be easy to see how to generalize the model. In what follows we specify the rules encountered at MRI in terms of hard and soft constraints.

J.O. Brunner, *Flexible Shift Planning in the Service Industry*, Lecture Notes in Economics and Mathematical Systems 640, DOI 10.1007/978-3-642-10517-3_3, © Springer-Verlag Berlin Heidelberg 2010

3.1.1.1 Hard Constraints

1. A shift must span a minimum length of time and can start at any time during a workday. The minimum shift length is defined by the hospital, and in our case is 6 consecutive hours without a break, or 7 h including an hour-long break.
2. A shift can be extended up to a maximum of 12 consecutive hours without a break assignment, or up to 13 h including a hour-long break, as specified in the general labor contract (see [88]).
3. After a shift or an on-call service (for a treatment of the latter see Sect. 3.2.1) ends a rest period of at least 12 h must follow [88]. Hence, a physician is not allowed to work two consecutive on-call services.
4. When breaks are required, each shift has to be assigned one (MRI does not schedule breaks but our full model will allow for them). In that case, the minimum and maximum shift lengths are extended by the length of the break. A break may start only after a fixed number of periods into a shift but has to start not later than a predefined set of periods worked and must end prior to a post-break workstretch of a fixed length.
5. A sufficient number of physicians must be on duty to cover the demand in each period.
6. A hospital-defined number of physicians must be available for each on-call service whose rules are stated by the hospital.
7. Regular demand does not occur on holidays, hence holiday periods are off days except in the case the physician is assigned to an on-call service.
8. If vacation time is permitted then the physician is off duty.

3.1.1.2 Soft Constraints

1. Each physician has an individual contract which specifies his or her regular working hours per week. All working time exceeding that amount is overtime. If fewer hours are assigned than are specified in the contract, the difference is undertime, which is first used to compensate overtime assigned in the previous week and then used for training and research if any remains.
2. Each physician can be assigned a maximum number of on-call services per week, which must be separated by at least one day. We assume in our model that the maximum number of on-call services per week is 1. The proscription of two consecutive services (hard constraint no. 3) is included in this assumption except at the interface of two consecutive weeks.
3. Considering each physician individually, all shift starting times in his or her schedule for an arbitrary week should be contained in an individual time window of pre-defined length.

Work requirements for each physician are spelled out in his or her contract and are currently set at a minimum of either 42, 48, or 54 regular hours per week. In particular, when the number of regular hours available is insufficient to meet demand, overtime can be assigned to avoid gaps in coverage. The use of overtime, though, is

3.1 Basic MIP Model

a costly option for the hospital so our goal is to try to match supply and demand as closely as possible.

Given the information above we can summarize the problem for different physician groups like full- and part-time physicians as follows. Suppose a set of physicians who are assigned to groups and who are to be scheduled over a planning horizon of $|\mathcal{T}|$ periods, normally stated in 1-h increments where $\mathcal{T}$ is the set of periods t. Each physician group (type) $j \in \mathcal{J}$ has certain defining characteristics, such as minimum shift length, maximum shift length, minimum rest duration between shifts, skill level, working time per week and overtime restrictions, as specified by general labor rules for each group or individual agreements between the physician groups and the hospital. Furthermore, it might be possible that a physician i within a group j has an individual agreement beyond the general or individual rules for his group with the hospital. For instance, the regular working time for a group is 42 h per week but one physician in this group would like to reduce his weekly workload to 34 h since he or she is close to retirement and does not need the money for full workload.

As it is typically the case, regular time can be viewed as a fixed cost determined by individual contracts. From that it follows that the relevant cost to be minimized is that which is paid out on top of the regular salary. The general labor contract which covers most of the physicians in Germany employs a complex rule to determine the payments. Only overtime from the current week that cannot be compensated by undertime in the following week is paid out. We call this special time *paid out time* to distinguish it form overtime. For example, if a physician works two hours of overtime in a week then the overtime hours has not to be paid out if the physician works at least two hours less than his normal working hours per week, say 42, in the subsequent week. In case the physician works one hour less which is 41 h then one hour of overtime has to be paid out. To this one hour we refer as paid out time. In any case, since it is not desirable to have schedules with overtime even if it can be compensated for we also minimize the overtime in each week of the planning horizon.

In addition, physicians assigned to shifts beyond a certain length must be given a break within some window. The minimum and maximum number of periods (work-stretch) before and the minimum number of periods after the break are treated as parameters. Physicians may also have a starting time window each day that may be absolute, relative, or a hybrid form of both over the week. An example of the latter would be the requirement that physicians of type j must be given rosters in which the latest start time minus the earliest start time within a week cannot exceed two hours and shifts cannot start before 10 a.m. To allow for an absolute time window a lower and upper bound is imposed on the earliest and latest start time variables, respectively.

During the week regular service is provided by physicians in sufficient numbers to meet the expected demand, $d_t, t \in \mathcal{T}^{dem} \subset \mathcal{T}$. During the evenings, nights, and on the weekends, on-call service is provided in fixed numbers, $d_t^{oc}, t \in \mathcal{L} \subset \mathcal{T}$, where $\mathcal{L}$ contains the set of starting periods for on-call services over the planning horizon. An on-call service in the hospital considered on a normal working day, Monday through Friday, follows a fixed 8-h shift, begins at 4 p.m., and lasts until 8 a.m. the next morning. On the weekends, Saturday and Sunday, only a 24-h on-call service needs to be scheduled starting at 8 a.m. on both days. Holidays are considered as weekend days. Following an on-call service, the physician cannot be assigned to a regular shift for some length of time. In other words, we have a minimum rest duration between shifts and/or on-call services. If physician group j has the skill to work on-call services, then the number of such services that each physician in the group can work in a week is limited. Furthermore, it is not allowed to assign a physician two consecutive on-call services since this would result in a 48-h on duty assignment.

When the house staff can meet only a portion of the demand in any period, our model schedules outside resources to fill in the gaps, but at a very high cost. Although it may be impractical to call on outside resources for short periods of time, their inclusion is more of a modeling device to guarantee feasibility and to alert management to the possible need for more staff. On a day-to-day basis, adjustments are made to the rosters by increasing overtime and adding part-timer hours in periods with higher than expected demand, or reassigning physicians when demand is low (see [17]). Other service organizations might have pool workers that can be used on short hand to fill gaps. For example, the German postal service has the possibility to call in students, housewives or retirees on an as needed basis.

To formulate the basic model, we need three sets of decision variables. The first signals whether or not a physician is on duty. The use of such variables is first introduced by Bowman [35] for a general schedule-sequencing problem. Let

$$
x_{j,i,t} = \begin{cases} 1, & \text{if group } j \in \mathcal{J} \text{ physician } i \in \mathcal{I}_j \text{ works in period } t \in \mathcal{T} \\ 0, & \text{otherwise,} \end{cases}
$$

where $\mathcal{I}_j$ is the set of physicians from group $j \in \mathcal{J}$ and, recall, $\mathcal{T}$ is the set of periods that spans the planning horizon. The next two variables that are at the heart of our new modeling approach indicate the period t when a shift begins for physician i, and when the corresponding rest period begins (cf. [98]). Let

$$
y_{j,i,t}^{shift} = \begin{cases} 1, & \text{if physician } i \in \mathcal{I}_j \text{ from group } j \text{ starts a shift} \\ & \text{at the beginning of period } t \in \mathcal{T} \\ 0, & \text{otherwise} \end{cases}
$$

$$
y_{j,i,t}^{rest} = \begin{cases} 1, & \text{if physician } i \in \mathcal{I}_j \text{ from group } j \text{ starts rest time} \\ & \text{at the beginning of period } t \in \mathcal{T} \\ 0, & \text{otherwise} \end{cases}
$$

3.1 Basic MIP Model

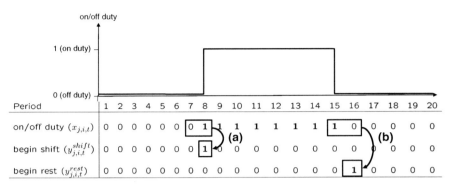

Fig. 3.1 Definition of variables

Figure 3.1 depicts an arbitrary assignment of the three variables over 20 periods for a physician along with the corresponding working profile. If we convert the graph into binary code we get the first row of the table below that is labeled with $x_{j,i,t}$. In this representation, a shift is identified by a string of 1s and a rest period by a string of 0s. In the example, the shift starts in period 8 so the variable $x_{j,i,8} = 1$ for physician i associated with group j and the variable $y^{shift}_{j,i,8} = 1$ as well. From a modeling point of view, this means if there is a transfer from a 0 in period $t-1$ to a 1 in the following period t for any decision variable $x_{j,i,t-1}$ and $x_{j,i,t}$ the connected variable $y^{shift}_{j,i,t}$ must also switch to 1 (see (a) in Fig. 3.1). When the shift ends, as it does at the end of period 15, the rest period must start in the next period. In this case, there is a transfer from a 1 to a 0 in the settings of the x-variables and a switch from 0 to 1 for the corresponding y^{rest} variable (see (b) in Fig. 3.1). This is shown in the bottom row of the Table in Fig. 3.1 where $y^{rest}_{j,i,16} = 1$. With these definitions, it is straightforward to introduce constraints that will ensure that each physician is given a feasible schedule over the planning horizon that considers minimum and maximum shift lengths as well as minimum rest time between consecutive shifts or other duty services.

3.1.2 Model Formulation

Again, the time horizon is divided into periods of fixed length, each 1-h long in the simplest case. For accounting purposes, the first period starts on Monday at 0:00 a.m. (midnight). Figure 3.2 shows the different indices we use in the modeling of the problem and illustrates the dependencies between indices and parameters. For instance, we disaggregate any day in 1 to T^{day} periods. Then a week consists of 7 days this translates to 1 to T^{week} periods within any week. Following this systematic the planning horizon can be divided evenly in $|\mathcal{W}|$ weeks which is equivalent to

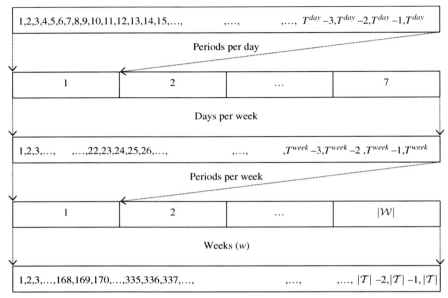

Fig. 3.2 Model indices

breaking into 1 to $|\mathcal{T}|$ periods of equal length. In the statement of the model we make use of the following notation.

Indices and sets

j	Index for physician groups
i	Index for physicians within a group
t	Index for periods
w	Index for weeks
$\mathcal{J}$	Set of physician groups
$\mathcal{I}_j$	Set of physicians within a group j
$\mathcal{T}$	Set of periods in planning horizon
$\mathcal{T}_w$	Subset of periods within a week w with
	$\mathcal{T}_w = \{t \in \mathcal{T} : (w-1) \cdot T^{week} < t \leq w \cdot T^{week}\} \subset \mathcal{T}$
$\mathcal{T}^{dem}$	Set of periods where demand occurs with
	$\mathcal{T}^{dem} = \{t \in \mathcal{T} : d_t > 0\} \subset \mathcal{T}$
$\mathcal{W}$	Set of weeks in planning horizon

3.1 Basic MIP Model

Parameters

d_t	Demand in period t
$c_{j,i}^{paid}$	Cost per hour of paid out time for physician i in group j
$c_{j,i}^{over}$	Cost per hour of overtime for physician i in group j
c^{out}	Cost per hour for an outside physician
$\overline{o}_{j,i}$	Maximal allowed overtime for a physician i in group j in a week
$\overline{T}_j^{shift}$	Maximum shift length for group j physicians
$\underline{T}_j^{shift}$	Minimum shift length for group j physicians
$\underline{T}_j^{rest}$	Minimum rest length after a shift ends for group j physicians
$r_{j,i}$	Regular working hours per week for physician i in group j according to general or individual agreement
T^{week}	Number of periods within a week; $T^{week} = |\mathcal{T}|/|\mathcal{W}|$
T^{day}	Number of periods within a day; $T^{day} = T^{week}/7$

Binary decision variables

$x_{j,i,t}$	1, If physician i associated with group j works in period t, 0 otherwise
$y_{j,i,t}^{shift}$	1, If physician i in group j begins a shift in period t, 0 otherwise
$y_{j,i,t}^{rest}$	1, If rest period begins for physician i in group j in period t, 0 otherwise

General integer decision variables

$o_{j,i,w}$	Amount of overtime for physician i in group j in week w
$u_{j,i,w}$	Amount of undertime for physician i in group j in week w
$h_{j,i,w}$	Amount of paid out time for physician i in group j in week w
x_t^{out}	Number of outside physicians hours hired in period t

Model

$$\text{Min} \sum_{w \in \mathcal{W}} \sum_{j \in \mathcal{J}} \sum_{i \in \mathcal{I}_j} \left(c_{j,i}^{paid} \cdot h_{j,i,w} + c_{j,i}^{over} \cdot o_{j,i,w} \right) + \sum_{t \in T^{dem}} c^{out} \cdot x_t^{out} \quad (1)$$

subject to

$$y_{j,i,t}^{shift} = x_{j,i,t} \cdot (1 - x_{j,i,t-1}), \forall j \in \mathcal{J}, i \in \mathcal{I}_j, t \in \mathcal{T} \tag{2}$$

$$y_{j,i,t}^{rest} = x_{j,i,t-1} \cdot (1 - x_{j,i,t}), \forall j \in \mathcal{J}, i \in \mathcal{I}_j, t \in \mathcal{T} \tag{3}$$

$$\sum_{\tau=t}^{\min(t+\underline{T}_j^{shift}-1,|\mathcal{T}|)} x_{j,i,\tau} \geq \min(\underline{T}_j^{shift}, |\mathcal{T}| - t + 1) \cdot y_{j,i,t}^{shift},$$
$$\forall j \in \mathcal{J}, i \in \mathcal{I}_j, t \in \mathcal{T} \tag{4}$$

$$\sum_{\tau=t}^{t+\overline{T}_j^{shift}} (1 - x_{j,i,\tau}) \geq y_{j,i,t}^{shift}, \forall j \in \mathcal{J}, i \in \mathcal{I}_j, t \in \{1, \ldots, |\mathcal{T}| - \overline{T}_j^{shift}\} \tag{5}$$

$$\sum_{\tau=t}^{\min(t+\underline{T}_j^{rest}-1,|\mathcal{T}|)} (1 - x_{j,i,\tau}) \geq \min(\underline{T}_j^{rest}, |\mathcal{T}| - t + 1) \cdot y_{j,i,t}^{rest},$$
$$\forall j \in \mathcal{J}, i \in \mathcal{I}_j, t \in \mathcal{T} \tag{6}$$

$$\sum_{j \in \mathcal{J}} \sum_{i \in \mathcal{I}_j} x_{j,i,t} + x_t^{out} \geq d_t, \forall t \in \mathcal{T}^{dem} \tag{7}$$

$$\sum_{t \in \mathcal{T}_w} x_{j,i,t} - o_{j,i,w} + u_{j,i,w} = r_{j,i}, \forall j \in \mathcal{J}, i \in \mathcal{I}_j, w \in \mathcal{W} \tag{8}$$

$$h_{j,i,w} \geq o_{j,i,w-1} - u_{j,i,w}, \forall j \in \mathcal{J}, i \in \mathcal{I}_j, w \in \mathcal{W} \tag{9}$$

$$x_{j,i,t} \text{ binary}, \forall j \in \mathcal{J}, i \in \mathcal{I}_j, t \in \mathcal{T} \cup \{0\} \tag{10}$$

$$y_{j,i,t}^{shift}, y_{j,i,t}^{rest} \text{ binary}, \forall j \in \mathcal{J}, i \in \mathcal{I}_j, t \in \mathcal{T} \tag{11}$$

$$x_t^{out}, \geq 0 \text{ and integer}, \forall j \in \mathcal{J}, i \in \mathcal{I}_j, t \in \mathcal{T}^{dem} \tag{12}$$

$$0 \leq o_{j,i,w} \leq \overline{o}_{j,i} \text{ and integer}, \forall j \in \mathcal{J}, i \in \mathcal{I}_j, w \in \mathcal{W} \cup \{0\} \tag{13}$$

$$u_{j,i,w}, h_{j,i,w} \geq 0 \text{ and integer}, \forall j \in \mathcal{J}, i \in \mathcal{I}_j, w \in \mathcal{W} \tag{14}$$

3.1 Basic MIP Model

The objective function (1) minimizes the sum of the paid out hours (i.e., the overtime from the previous week that could not be compensated by undertime in the current week), the costs of overtime, and the costs for outside physician hours. Implicitly, the goal is to cover demand with regular hours first, then with overtime, and finally with outside hours when the first two options are exhausted. Overtime does not incur real costs in the final schedules as long as it can be compensated for by undertime in the following week. To achieve the desired results, the cost coefficients in (1) must be defined in relative terms, as will be discussed shortly.

The first block of constraints (2)–(3) defines the beginning and the end of a shift, and by implication, the start of a rest period. Constraints (2) ensure that if a shift begins in period t, then the variable $y_{j,i,t}^{shift}$ is 1. In contrast, constraints (3) force variable $y_{j,i,t}^{rest}$ to 1 if a shift ends in period $t-1$ (see Fig. 3.1 for a definition of the variables). Note that constraints (2) and (3) are nonlinear but can be linearized with standard techniques. The procedure used to linearize them is given as follows (cf. [121]). Applying the linearization constraints (2) and (3) are replaced by

$$-x_{j,i,t} + y_{j,i,t}^{shift} \leq 0, \forall j \in \mathcal{J}, i \in \mathcal{I}_j, t \in \mathcal{T} \tag{2a}$$

$$x_{j,i,t-1} + y_{j,i,t}^{shift} \leq 1, \forall j \in \mathcal{J}, i \in \mathcal{I}_j, t \in \mathcal{T} \tag{2b}$$

$$x_{j,i,t} - x_{j,i,t-1} - y_{j,i,t}^{shift} \leq 0, \forall j \in \mathcal{J}, i \in \mathcal{I}_j, t \in \mathcal{T} \tag{2c}$$

and

$$-x_{j,i,t-1} + y_{j,i,t}^{rest} \leq 0, \forall j \in \mathcal{J}, i \in \mathcal{I}_j, t \in \mathcal{T} \tag{3a}$$

$$x_{j,i,t} + y_{j,i,t}^{rest} \leq 1, \forall j \in \mathcal{J}, i \in \mathcal{I}_j, t \in \mathcal{T} \tag{3b}$$

$$x_{j,i,t-1} - x_{j,i,t} - y_{j,i,t}^{rest} \leq 0, \forall j \in \mathcal{J}, i \in \mathcal{I}_j, t \in \mathcal{T} \tag{3c}$$

respectively.

The second block of constraints (4)–(6) defines the interaction between possible shift lengths and rest time. For each feasible starting period t, the minimum shift length $\underline{T}_j^{shift}$ defined for group j is enforced by constraints (4) and the maximum shift length $\overline{T}_j^{shift}$ by constraints (5). More specifically, if a shift starts in period t, constraints (4) force the x's to be 1 for t up to $t + \underline{T}_j^{shift} - 1$, while constraints (5) ensure that at least one of the $x_{j,i,\tau}$'s from $\tau = t$ to $t + \overline{T}_j^{shift}$ is set to 0 and hence the shift is ended. A minimum amount of rest between consecutive shifts is achieved with constraints (6). Taking up our example from Fig. 3.1 we show exemplary in Fig. 3.3 how constraints (4) and (6) work. Recall, a shift starts in period

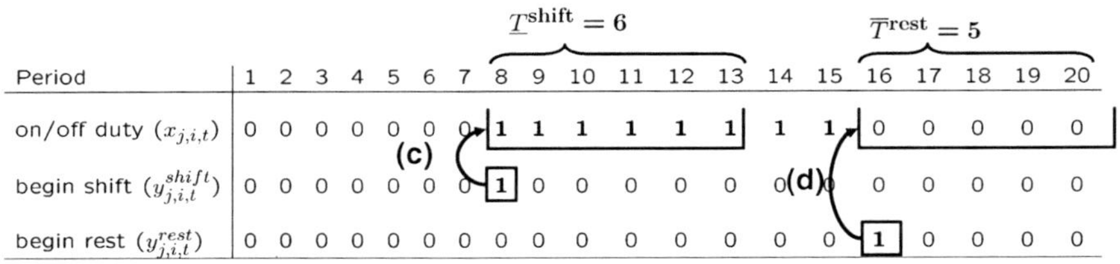

Period	1	2	3	4	5	6	7	8	9	10	11	12	13	14	15	16	17	18	19	20
on/off duty $(x_{j,i,t})$	0	0	0	0	0	0	0	1	1	1	1	1	1	1	1	0	0	0	0	0
begin shift $(y_{j,i,t}^{shift})$	0	0	0	0	0	0	0	1	0	0	0	0	0	0	0	0	0	0	0	0
begin rest $(y_{j,i,t}^{rest})$	0	0	0	0	0	0	0	0	0	0	0	0	0	0	0	1	0	0	0	0

With overbraces: $\underline{T}^{shift} = 6$ over periods 8–13; $\overline{T}^{rest} = 5$ over periods 16–20. Annotations **(c)** and **(d)**.

Fig. 3.3 Definition of shift design constraints

8 and ends after period 15. Constraints (4) forces that the physician is on duty for at least $\underline{T}_j^{shift} = 6$ periods after period 8 inclusively (see (c) in Fig. 3.3). Almost the same holds for a rest period except that there is a minimum length where the physician is off duty, in our case five periods, that is $\underline{T}_j^{rest} = 5$ (see (d) in Fig. 3.3). Note that the maximum shift length is not depicted in Fig. 3.3.

Constraints (7) ensure that the net workforce is sufficient to cover the demand d_t for each period $t \in \mathcal{T}^{dem}$ in the planning horizon where $\mathcal{T}^{dem}$ is the set of all positive demand periods, $d_t > 0 \; \forall t \in \mathcal{T}$. To guarantee that the model is feasible when demand exceeds the number of available physicians on any day in the planning horizon, we include for every period the possibility to hire outside physician hours for an arbitrary amount with the integer variable x_t^{out} and define their objective function coefficient to be $c^{out} > \sum_{j \in \mathcal{J}} \sum_{i \in \mathcal{I}_j} \overline{o}_{j,i} \cdot c_{j,i}^{over}$.

The third block of constraints (8)–(9) serves to determine the number of paid out hours in each week. Recall that the paid out time is the overtime from the previous week that is not compensated by undertime in the current week. Equation (8) calculate the amount of over- or undertime with respect to regular working hours, $r_{j,i}$, for each physician i associated with group j in week w. The summation counts the hours worked in each week by regular shift assignments. Constraints (9) calculate the paid out hours for a week $w \in W$. When a rolling time horizon is used initial conditions must be provided for $o_{j,i,0}$ (this is further discussed in Sect. 4.1). To forbid the occurrence of over- and undertime in the same week the cost coefficients in the objective function have to satisfy $c_{j,i}^{over} > c_{j,i}^{paid}$ for all on-staff physicians $i \in \mathcal{I}_j$ corresponding to group $j \in \mathcal{J}$.

Variable definitions are given in (10)–(14); however, the binary conditions on $y_{j,i,t}^{shift}$ and $y_{j,i,t}^{rest}$ can be relaxed because constraints (2) and (3) force them to binary values since the x variables are binary. Also, the integral requirements on x_t^{out}, $h_{j,i,w}$, $o_{j,i,w}$ and $u_{j,i,w}$ can be relaxed because they will automatically be satisfied in any feasible solution. For x_t^{out} and $h_{j,i,w}$, this is a direct result of constraints (7) and (9), and the fact that we are minimizing a linear function of these variables in (1). Because the x variables are binary and the demand is integral, (1) and (7) imply that x_t^{out} is integral too. From (9) we have $h_{j,i,w} = \min\{0, o_{j,i,w-1} - u_{j,i,w}\}$ so $h_{j,i,w}$ is integral when $o_{j,i,w}$ and $u_{j,i,w}$ are integral. For the latter two variables, the cost coefficients in the objective function imply that at most one of them can be positive in any feasible solution. Now, given that $x_{j,i,t}$ and $r_{j,i}$ are integral in (8), we can conclude the same for $o_{j,i,w}$ and $u_{j,i,w}$ and hence for $h_{j,i,w}$.

3.2 Model Enhancements

A solution to model (1)–(14) provides a basic work schedule over the planning horizon: it accommodates the general rules with respect to minimum and maximum shift lengths and it guarantees sufficient rest between shift assignments. Another basic feature of physician scheduling in a hospital is considering on-call services that has to be incorporated to cover off hours where emergencies might occur. Furthermore, we have found that additional constraints are needed to obtain high quality rosters. For instance, physicians prefer not to start a shift early one day and late the next. Moreover, there are rules specified in the general labor contract for break assignments that should be taken into account (see [88]). Finally, accommodation of holidays and vacation requests in the scheduling procedure is crucial for an application in a hospital. We address these issues below, first presenting a formulation for on-call services, then introducing time window restrictions on start times for shifts and a set of break constraints to accommodate lunch breaks, and finally showing how to incorporate holidays and vacation requests in the MIP formulation.

3.2.1 On-Call Services

We introduce a new set of decision variables that is associated with the on-call assignments that are needed to provide coverage during off-hours, typically nights and weekends. The number of physicians that are required for this purpose could be estimated from past data but it is more common for hospital administrators to fix the requirements based on experience. The goal is to provide a sufficient number of physicians to cover the random demand in all but extreme situations. For MRI, on-call service begins at 4 p.m. on weekdays (Monday–Friday) and lasts 16 h. Before that service begins on weekdays the assigned physician has to work a regular fixed 8-h shift, thus the combined service starts at 8 a.m. on Monday through Friday and lasts 24 h. On Saturday and Sunday, no regular shifts are scheduled because there are no planned operations; on-call service begins at 8 a.m. and lasts 24 h. We refer to both the combined on-call service and fixed shift on weekdays and the pure on-call service on the weekends as *on-call service* because their structure with respect to starting time and duration is the same. This simplifies our formulation because each on-call service can be modeled as a 24-h assignment. The only necessary consideration when modeling both on-call services as a 24-h service is that the fixed 8-h shift on weekdays, before the pure on-call service starts, affects the demand profile (see Sect. 4.1).

After any on-call service, a physician must have a minimum number of hours off during which no shift is allowed to start. Moreover, the number of on-call services that can be assigned throughout the week is limited. To accommodate the requirements for on-call services, we define a set $\mathcal{L}$ of possible time periods in which on-call services can start that include both weekdays and weekends. One additional point that has to be considered is that only a fraction of the worked hours during an on-call service is counted toward regular working hours; the remaining hours are paid out. However, a portion of the hours worked on a Sunday spill over to Monday and so count towards regular hours in the next week.

The following accounting scheme applies. For an on-call service on Monday–Thursday, 16 of the 24 h are regular hours and the remaining 8 h are paid out. On Friday, 8 of the 24 h are added to regular hours and the remaining 16 h are paid out. On Saturday, all 24 h are paid out; on Sunday, 16 of the 24 h are paid out and the remaining 8 h count towards regular working hours in the following week. To account for the start of an on-call service, we use a variable similar to the one used for starting a shift. Let

$$
y_{i,t}^{oc} = \begin{cases} 1, & \text{if physician } i \text{ associated with group } j \text{ begins an on-call} \\ & \text{service at the beginning of period } t \in \mathcal{L} \\ 0, & \text{otherwise} \end{cases}
$$

To plan for on-call services in the rostering process additional notation is used.

Indices and sets

l Index for periods

$\mathcal{L}$ Set of starting periods for an on-call service

$\mathcal{L}_w$ Subset of starting periods for an on-call service within a week w with

$$\mathcal{L}_w = \{t \in \mathcal{L} : (w-1) \cdot T^{week} < t \leq w \cdot T^{week}\} \subset \mathcal{L}$$

Parameters

c^{oc_out} Cost for an outside physician who conducts an entire on-call service

T^{oc} Length of an on-call service

d_t^{oc} Number of physicians required for each on-call service

           starting in period $t \in \mathcal{L}$

$\overline{n}_j^{oc}$ Maximum number of on-call services for a group j physician in any week

Function

$f_1(t)$ Calculates the number of hours that are charged for any on-call service

           for each $t \in \mathcal{L}$ to regular working time per week

Binary decision variable

$y_{j,i,t}^{oc}$ 1, If j, i begins on-call service in $t \in \mathcal{L}$, 0 otherwise

3.2 Model Enhancements

General integer decision variable

$x_t^{oc\_out}$ Number of outside physicians who carry out an on-call service
starting in period $t \in \mathcal{L}$

In what follows we state the new constraints to accommodate on-call services in the formulation.

$$\sum_{\tau=\max(t-\underline{T}_j^{rest},0)}^{t-1} (1 - x_{j,i,\tau}) \geq \min(\underline{T}_j^{rest}, t-1) \cdot y_{j,i,t}^{oc}, \tag{15}$$

$$\forall j \in \mathcal{J}, i \in \mathcal{I}_j, t \in \mathcal{L}$$

$$\sum_{\tau=t}^{\min(t+\underline{T}_j^{rest}+T^{oc}-1,|\mathcal{T}|)} (1 - x_{j,i,\tau}) \geq \min(\underline{T}_j^{rest} + T^{oc}, |\mathcal{T}| - t + 1) \cdot y_{j,i,t}^{oc}, \tag{16}$$

$$\forall j \in \mathcal{J}, i \in \mathcal{I}_j, t \in \mathcal{L}$$

$$\sum_{j \in \mathcal{J}} \sum_{i \in \mathcal{I}_j} y_{j,i,t}^{oc} + x_t^{oc\_out} \geq d_t^{oc}, \forall t \in \mathcal{L} \tag{17}$$

$$\sum_{t \in \mathcal{L}_w} y_{j,i,t}^{oc} \leq \overline{n}_j^{oc}, \forall j \in \mathcal{J}, i \in \mathcal{I}_j, w \in \mathcal{W} \tag{18}$$

$$y_{j,i,t}^{oc} + y_{j,i,t+T^{day}}^{oc} \leq 1, \forall j \in \mathcal{J}, i \in \mathcal{I}_j, t \in \left\{ l \in \mathcal{L} : l < |\mathcal{T}| - T^{day} \right\} \tag{19}$$

$$y_{j,i,t}^{oc} \text{ binary}, x_t^{oc\_out} \geq 0 \text{ and integer}, \forall j \in \mathcal{J}, i \in \mathcal{I}_j, t \in \mathcal{L} \tag{20}$$

Constraints (15)–(20) ensure that the necessary number of on-call services are generated over the planning horizon. In detail, constraints (15) require that at least $\underline{T}_j^{rest}$ rest periods are included in the line-of-work of physician i associated with group j when he or she is assigned an on-call service starting in period $t \in \mathcal{L}$. In contrast, constraints (16) ensure that if a physician is scheduled to an on-call service starting in period $t \in \mathcal{L}$, the physician is on duty for the next T^{oc} periods (24 h in our case) and then off duty for at least $\underline{T}_j^{rest}$ periods following the service according to its labor contract but the regular 8-h shift included in an on-call service on Monday–Friday is considered in a preprocessing step (see Sect. 4.1). Note, that if a physician is assigned an on-call service starting in period $t \in \mathcal{L}$ we force the x's to 0 for periods t up to $t + T^{oc} + \underline{T}_j^{rest}$ in his roster meaning that the physician cannot be assigned a regular shift in this period of time.

Constraints (17) force the number of physicians who are assigned to an on-call service starting in period $t \in \mathcal{L}$ to be at least d_t^{oc}. Normally, the hospital schedules exactly d_t^{oc} physicians to an on-call service. However, we included the possibility to hire an outside physician to conduct the on-call service and hence relax the equality constraint to an covering constraint. In case more than d_t^{oc} physicians are assigned a simple post processing step is carried out which deletes redundant physicians and recalculates the cost associated with their final roster. To prohibit a solution where outside resources are used to cover on-call services we weight the usage of these resources in the objective function (21) with large cost c^{oc_out}. Recall, that we penalize also the use of outside hours to cover regular demand periods d_t. Constraints (18) permit at most $\bar{n}_j^{oc}$ services to be assigned to any physician associated with group j in each week and (19) forbid the occurrence of two consecutive on-call services for a physician. Again, initial conditions must be considered for the last on-call assignments of the previous schedule when a rolling time horizon is considered (this is further discussed in Sect. 4.1). Constraints (20) give the variable definition where integrality conditions can be ignored for $x_t^{oc\_out}$. The same argument as for x_t^{out} is employed.

As mentioned, the objective function (1) changes to (21) to consider the case when outside recourses are necessary to cover an on-call service starting in period $t \in \mathcal{L}$.

$$\text{Min} \sum_{w \in \mathcal{W}} \sum_{j \in \mathcal{J}} \sum_{i \in \mathcal{I}_j} \left(c_{j,i}^{paid} \cdot h_{j,i,w} + c_{j,i}^{over} \cdot o_{j,i,w} \right)$$
$$+ \sum_{t \in \mathcal{T}^{dem}} c^{out} \cdot x_t^{out} + \sum_{t \in \mathcal{L}} c^{oc\_out} \cdot x_t^{oc\_out} \quad (21)$$

Additionally, to the new constraints (15)–(20) and the extension of the objective function (1)–(21) we have to modify constraints (8) to take consideration of on-call services into account when determining the workload in any week and hence the over- and undertime.

$$\sum_{t \in \mathcal{T}_w} x_{j,i,t} + \sum_{t \in \{l \in \mathcal{L}_w : l > T^{day}\}} f_1(t - T^{day}) \cdot y_{j,i,t-T^{day}}^{oc}$$
$$= r_{j,i} + o_{j,i,w} - u_{j,i,w}, \forall j \in \mathcal{J}, i \in \mathcal{I}_j, w \in \mathcal{W} \quad (22)$$

The new constraints (22) calculate the amount of over- or undertime with respect to regular working hours, $r_{j,i}$, for each physician i associated with group j in week w. Again, the first summation counts the hours worked in each week by regular shifts, whereas the second summation adds a portion of the hours rendered during an on-call service, if assigned to the physician i associated to group j. Hence, the function f_1 allocates the exact portion of the hours worked during an on-call service to the regular working hours for the week. The rules stated above are applied to define the function f_1. The set of on-call services within a week, $\mathcal{L}_w$, contains the starting periods for each service within a week (for MRI period 9, 33, 57 and so forth). For instance, if a physician is assigned to an on-call service on Wednesday in the first week of the planning horizon, which corresponds to $t = 57$ in the set $\mathcal{L}_w$, the function $f_1(57) = 16$ states, that 16 h are counted to regular hours in the week. By inference, then, if a physician is assigned an on-call service on any day of the week, he does not work the next day.

3.2.2 Time Window Restrictions

In the context of staff scheduling, a time window is a set of consecutive periods of pre-defined length during which a shift is allowed to start. Individual staff members may have different time windows. For modeling purposes we introduce the following additional notation.

Set, parameter, and function

$\mathcal{S}_w$ Set of first periods within any day in a week w with

$$\mathcal{S}_w = \{t \in \mathcal{T}_w : t \bmod T^{day} = 1\} \subset \mathcal{T}_w$$

T_j^{win} Length of starting time window for physician group j

$f_2(t)$ Maps each period $t \in \mathcal{T}$ into a period during a specific day, where

$$\mathcal{T} \longmapsto \{1, \ldots, T^{day}\} \text{ with } t \mapsto f_2(t) = t \bmod T^{day}$$

Decision variables

$e_{j,i,w}$ Earliest starting time for physician i in group j in week w

$l_{j,i,w}$ Latest starting time for physician i in group j in week w

We now present the following new constraints to accommodate time windows in the formulation.

$$e_{j,i,w} \leq T^{day} - \sum_{\tau=t}^{t+T^{day}-1} (T^{day} - f_2(\tau)) \cdot y_{j,i,\tau}^{shift}, \tag{23}$$
$$\forall j \in \mathcal{J}, i \in \mathcal{I}_j, w \in W, t \in \mathcal{S}_w$$

$$l_{j,i,w} \geq \sum_{\tau=t}^{t+T^{day}-1} f_2(\tau) \cdot y_{j,i,\tau}^{shift}, \forall j \in \mathcal{J}, i \in \mathcal{I}_j, w \in W, t \in \mathcal{S}_w \tag{24}$$

$$l_{j,i,w} - e_{j,i,w} \leq T_j^{win}, \forall j \in \mathcal{J}, i \in \mathcal{I}_j, w \in W \tag{25}$$

$$l_{j,i,w}, \; e_{j,i,w} \geq 0 \text{ and integer}, \forall j \in \mathcal{J}, i \in \mathcal{I}_j, w \in W \tag{26}$$

Constraints (23) and (24) calculate the earliest and latest shift starting time for any full week and each physician, respectively. The function $f_2(t)$ maps the periods $t \in \mathcal{T}$ to periods within a day; for example, if $t = 25$ then $f_2(25) = 1$ which means that $t = 25$ corresponds to the first period on the second day. We sum over T^{day} periods to identify the starting time for each day. In our application, the data implicitly limits the number of shifts for any physician on a day to at most one [see Sect. 3.3]. The next constraints (25) bound the difference between the latest and earliest start times, $l_{j,i,w} - e_{j,i,w}$. For any physician i associated with group j, his shifts are not allowed to start more than T_j^{win} periods apart. Because the model forces this difference to be as tight as necessary and the fact that $l_{j,i,w}$ is maximized [see Constraints (24)] and $e_{j,i,w}$ is minimized [see Constraints (23)] these variables will always be integer in any feasible solution so the integrality requirement in constraints (26) can be relaxed. Note, the right-hand sides of constraints (23) and (24) are general integer which sets $l_{j,i,w}$ and $e_{j,i,w}$ to integer values.

3.2.3 Break Assignment

There are several different ways of accommodating breaks in the formulation, depending on the restrictions that the hospital or department manager wishes to impose. Note that regular shifts which are part of an on-call service (see Sect. 3.2.1) have a predetermined break time (cf. alternative (1.) below). In this case, since on-call services have a fixed starting time breaks are not modeled via constraints but are taken into account by preprocessing (see Sect. 4.1). The following three alternatives consider the inclusion of breaks for regular shifts into the formulation.

1. The break takes place in a predefined time span after the shift starts.
2. The break must be located in a time interval that begins after a minimum pre-break workstretch and ends before a minimum post-break workstretch. In this case, a physician must work a minimum and maximum amount of time before and after the break is assigned.
3. Only shifts that are longer than a specific amount of time require a break.

3.2.3.1 Break Placement After a Predefined Time Span

Assuming that the length of a break is equal to a period, the first possibility (1.) is the simplest to implement because no additional constraints or variables are needed. However, we do need a new parameter for each group j physicians that fixes the break for any shift to a specific period following its start. Let T_j^{fix} be the fixed

3.2 Model Enhancements

number of periods that must be worked before a break begins, and observe that both the demand constraints (7) and the working time constraints (8) will be affected when a period is removed from a shift. To identify the break assignment period, we use the variable $y_{j,i,t}^{shift}$ and modify the two sets of constraints for the periods in the set $\{t \in \mathcal{T} : t > T_j^{fix}\}$. For all other periods in $\mathcal{T}$ the constraint is built according to the original formulation [see (7)].

$$\sum_{j \in \mathcal{J}} \sum_{i \in \mathcal{I}_j} (x_{j,i,t} - y_{j,i,t-T_j^{fix}}^{shift}) + x_t^{out} \geq d_t, \forall t \in \mathcal{T}^{dem} \tag{7a}$$

$$\sum_{t \in \mathcal{T}_w} x_{j,i,t} - \sum_{t \in \mathcal{T}_w \setminus \{1,\ldots,T_j^{fix}\}} y_{j,i,t-T_{j,i}^{fix}}^{shift} + \sum_{t \in \{l \in \mathcal{L}_w : l > T^{day}\}} f_1(t - T^{day}) \cdot y_{j,i,t-T^{day}}^{oc}$$

$$= r_{j,i} + o_{j,i,w} - u_{j,i,w}, \forall j \in \mathcal{J}, i \in \mathcal{I}_j, w \in \mathcal{W} \tag{8a}$$

3.2.3.2 Implicit Break Placement in a Defined Time Interval

The second alternative offers more flexibility than the first but still requires that every shift must be given a break. Here, a break may not start before a minimum pre-break amount of time $\underline{T}_j^{pre}$ is worked and must be assigned before a maximum pre-break amount of time $\overline{T}_j^{pre}$ is worked. Furthermore, the break must be assigned before the shift ends minus a post-break number of periods $\underline{T}_j^{post}$. For modeling purposes, a new variable is needed. Let

$$y_{j,i,t}^{break} = \begin{cases} 1, & \text{if physician } i \in \mathcal{I}_j \text{ from group } j \text{ has a break in period } t \in \mathcal{T} \\ 0, & \text{otherwise.} \end{cases}$$

We now have to introduce the following new constraints.

$$\sum_{\tau=t+\underline{T}_j^{pre}}^{\min\{t+\overline{T}_j^{pre}, |\mathcal{T}|\}} y_{j,i,\tau}^{break} \geq y_{j,i,t}^{shift}, \forall j \in \mathcal{J}, i \in \mathcal{I}_j, t \in \mathcal{T} \setminus \{|\mathcal{T}| - \underline{T}_j^{pre} + 1, \ldots, |\mathcal{T}|\} \tag{27}$$

$$\sum_{\tau=t-\overline{T}_j^{shift}+\underline{T}_j^{pre}}^{t-\underline{T}_j^{post}-1} y_{j,i,\tau}^{break} \geq y_{j,i,t}^{rest}, \forall j \in \mathcal{J}, i \in \mathcal{I}_j, t \in \{\overline{T}_j^{shift} - \underline{T}_j^{pre}, \ldots, |\mathcal{T}|\} \tag{28}$$

$$\sum_{t \in \mathcal{T}} y_{j,i,t}^{shift} = \sum_{t \in \mathcal{T}} y_{j,i,t}^{break}, \forall j \in \mathcal{J}, i \in \mathcal{I}_j \tag{29}$$

$$y_{j,i,t}^{break} \leq x_{j,i,t}, \forall j \in \mathcal{J}, i \in \mathcal{I}_j \tag{30}$$

$$y_{j,i,t}^{break} \text{ binary}, \forall j \in \mathcal{J}, i \in \mathcal{I}_j, t \in \mathcal{T} \tag{31}$$

Constraints (27) force a break following each shift start but not until $\underline{T}_j^{pre}$ periods have been worked. Constraints (28) ensure that a break is assigned at least $\underline{T}_j^{post}$ periods before the end of the respective shift. Finally, constraints (29) state that the number of breaks must be equal to the number of shift starts for each physician i and constraints (30) force that if a break assignment takes place then the corresponding on duty variable has to be 1 too. Variable definition is given in (31). To complete the formulation, constraints (7) and (8) are straightforward modified as follows.

$$\sum_{j \in \mathcal{J}} \sum_{i \in \mathcal{I}_j} \left(x_{j,i,t} - y_{j,i,t}^{break} \right) + x_t^{out} \geq d_t, \forall t \in \mathcal{T}^{dem} \tag{7b}$$

$$\sum_{t \in \mathcal{T}_w} (x_{j,i,t} - y_{j,i,t}^{break}) - o_{j,i,w} + u_{j,i,w} = r_{j,i}, \forall j \in \mathcal{J}, i \in \mathcal{I}_j, w \in \mathcal{W} \tag{8b}$$

Figure 3.4 depicts the assignment of variables shown in Fig. 3.1 with the exception that break assignments are considered too. Again, in this representation, a shift

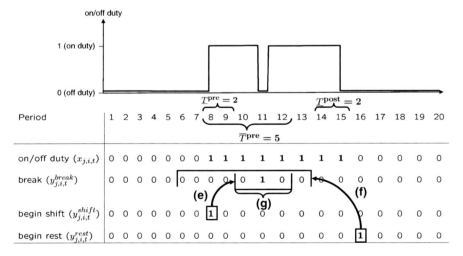

Fig. 3.4 Definition of variables including break assignments

3.2 Model Enhancements

is identified by a string of 1s and a rest period by a string of 0s. In Fig. 3.4, the shift starts in period 8 and ends after period 15. We introduce a new row that signals whether period t is a break period or not. Now by using the second alternative to define breaks and assuming that the physician has to work two periods before $\left(\underline{T}_j^{pre} = 2\right)$ and after $\left(\underline{T}_j^{post} = 2\right)$ a break assignment takes place as well as that the break has to be within five periods following a shift start $\left(\overline{T}_j^{pre} = 5\right)$ the range of possible break assignment periods (g) is from period 10 to 12. Corresponding periods are highlighted in the period column in Fig. 3.4. Constraints (27) take the form (d) whereas constraints (28) are presented as (e). Consequently one of the three $y_{j,i,t}^{break}$ variables has to be 1. The three variables are determined by the overlapping interval of constraints (27) and (28). In our case the break is assigned to period 11 and hence $y_{j,i,11}^{break} = 1$. The upper part of Fig. 3.4 shows the availability of the physician. As can be seen, the physician is off duty in period 11. We model the availability in the formulation as $x_{j,i,t} - y_{j,i,t}^{break}$ [see constraints (7b)]. With the additional definitions, it is straightforward to incorporate breaks in the planning process.

3.2.3.3 Break Placements When Short Shifts are Considered

The third possibility (3.) is very similar to the second, except that now shifts that are shorter than T_j^{short} periods are not required to have a break. Although this relaxation allows for more flexibility, it makes the model more complex. To formulate the necessary constraints, we introduce the following binary variable

$$
y_{j,i,t}^{short} = \begin{cases} 1, & \text{if physician } i \text{ associated with group } j \text{ works a shift} \\ & \text{shorter than } T_j^{short} \text{ starting in period } t \\ 0, & \text{otherwise} \end{cases}
$$

and replace constraints (27)–(29) by (32)–(34). In addition, three new sets of constraints (35)–(37) are needed to identify the short shifts. The constraints (7b) and (8b) remain the same.

$$
\sum_{\tau=t+\underline{T}_j^{pre}}^{\min\{t+\overline{T}_j^{pre},|\mathcal{T}|\}} y_{j,i,\tau}^{break} \geq y_{j,i,t}^{shift} - y_{j,i,t}^{short}, \tag{32}
$$

$$
\forall j \in \mathcal{J}, i \in \mathcal{I}_j, t \in \{1, \ldots, |\mathcal{T}| - \underline{T}_j^{pre} + 1\}
$$

$$\sum_{\tau=t-\overline{T}_j^{shift}+\underline{T}_j^{pre}}^{t-\underline{T}_j^{post}-1} y_{j,i,\tau}^{break} \geq y_{j,i,t}^{rest} - \sum_{\tau=t-T_j^{short}}^{t-1-\underline{T}_j^{shift}} y_{j,i,\tau}^{short}, \tag{33}$$

$$\forall j \in \mathcal{J}, i \in \mathcal{I}_j, t \in \{\overline{T}_j^{shift} - \underline{T}_j^{pre}, \dots, |\mathcal{T}|\}$$

$$\sum_{t \in \mathcal{T}}(y_{j,i,t}^{shift} - y_{j,i,t}^{short}) = \sum_{t \in \mathcal{T}} y_{j,i,t}^{break}, \forall j \in \mathcal{J}, i \in \mathcal{I}_j \tag{34}$$

Constraints (32) force the assignment of a break to a shift after a pre-break amount of time but only in the case when the shift is of required length. The same logic is applied by constraints (33) but for the post-break case. Constraints (34) determine the number of breaks that have to be assigned. The two constraints, (35) and (36), determine whether or not a shift is a short shift. The last constraints (37) ensure that the indicator variable $y_{j,i,t}^{short}$ is 1 only when the corresponding shift starting variable $y_{j,i,t}^{shift}$ is 1. Variable definitions are given in (38).

$$\sum_{\tau=t}^{\min\{t+\overline{T}_j^{shift},|\mathcal{T}|\}} (\tau \cdot y_{j,i,\tau}^{rest}) - t \cdot y_{j,i,t}^{shift} \tag{35}$$
$$\geq (T_j^{short} + 1) \cdot \left(y_{j,i,t}^{shift} - y_{j,i,t}^{short}\right), \forall j \in \mathcal{J}, i \in \mathcal{I}_j, t \in \mathcal{T}$$

$$\sum_{\tau=t}^{\min\{t+\overline{T}_j^{shift},|\mathcal{T}|\}} (\tau \cdot y_{j,i,\tau}^{rest}) - t \cdot y_{j,i,t}^{shift} \tag{36}$$
$$\leq T_j^{short} \cdot y_{j,i,t}^{shift} + |\mathcal{T}| \cdot (1 - y_{j,i,t}^{short}), \forall j \in \mathcal{J}, i \in \mathcal{I}_j, t \in \mathcal{T}$$

$$y_{j,i,t}^{shift} \geq y_{j,i,t}^{short}, \forall j \in \mathcal{J}, i \in \mathcal{I}_j, t \in \mathcal{T} \tag{37}$$

$$y_{j,i,t}^{break}, y_{j,i,t}^{short} \text{ binary}, \forall j \in \mathcal{J}, i \in \mathcal{I}_j, t \in \mathcal{T} \tag{38}$$

3.2.4 Holidays and Vacations

Another feature not included in the basic model (1)–(14) concerns holidays during which no regular time is scheduled. Let $\mathcal{H} \subset \mathcal{T}$ be the set of holiday periods other than weekends, where $d_t = 0, \forall t \in \mathcal{H}$. The following constraints ensure that no rosters will be generated that cover $t \in \mathcal{H}$ with regular time, although on-call assignments are permissible during these periods.

3.2 Model Enhancements

$$\sum_{j \in \mathcal{J}} \sum_{i \in \mathcal{I}_j} \sum_{t \in \mathcal{H}} x_{j,i,t} \leq 0 \qquad (39)$$

The left-hand side of constraints (39) counts all x-variables corresponding to periods on holidays for all physicians $i \in \mathcal{I}_j$ and physician types $j \in \mathcal{J}$. This term is then set to ≤ 0 to forbid regular shift assignments on holidays.

A similar idea can be applied to count for vacation request by a physician i of type j. Let $\mathcal{V}_{j,i} \subset \mathcal{T}$ be the set of vacation periods and $\mathcal{V}_{j,i}^{day} \subset \mathcal{L}$ be the corresponding set of vacation days in the planning horizon where an element is associated with exact one element in $\mathcal{L}$ for physician $i \in \mathcal{I}_j$ associated with type $j \in \mathcal{J}$. Then the following constraints assure that the corresponding physicians are off duty when vacation is permitted.

$$\sum_{t \in \mathcal{V}_{j,i}} x_{j,i,t} + \sum_{t \in \mathcal{V}_{j,i}^{day}} y_{j,i,t}^{oc} \leq 0, \forall j \in \mathcal{J}, i \in \mathcal{I}_j \qquad (40)$$

Note that the physician cannot be assigned to an on-call service as it is the case for holidays. The first term in constraints (40) contains all regular on duty variables (x-variables) for a specific physician $i \in \mathcal{I}_j$ during the period of vacation whereas the second term counts the corresponding on-call starting variables for that period. The sum of both terms is set to less than or equal to 0 and hence no assignment can take place in the period of a permitted vacation since decision variables are subject to the non-negativity restriction.

Other considerations of a preferential nature are more or less straightforward to implement. For example, if a physician prefers not to start his or her shift before 10 a.m., all that is needed is to set a lower bound on the earliest starting time variable $e_{j,i,w}$ to 11. Alternatively, to forbid a shift from starting after a specific period, the latest starting time variable $l_{j,i,w}$ can be bounded accordingly for the respective physician. Balancing overtime hours or on-call services during a week, however, would require the introduction of additional variables and constraints. It is possible to model these considerations as hard constraints, or as soft constraints using goal programming techniques (e.g. see [105]). Goal programming is a popular modeling technique to consider more objectives rather than just minimizing costs. In general, goals or target values are defined for some constraints and the deviation from that target values is penalized in the objective function. For example, a nurse specifies the goal that she wants to work two night shifts each week if possible but she can also work one or three night shifts per week. In the latter case the deviation of 1 would be penalized in the cost function. From a computational perspective goal programming introduces more variables (lower and upper deviation variables) to the model. Additionally, the optimization problem is a multiple criteria decision making methodology with lexicographic weights for deviation variables (deviation from target values). From a application point of view the target values has to be specified exactly. Also, the weights has to be determined which might be a complex task too, especially when the order of the goals is not obvious or when the goals are defined in different units. To specify the weights analytic hierarchy process techniques can be used (e.g. see [33]). The analytic hierarchy process is a technique that helps the decision makers to find the weights that fit best to their needs rather then prescribing correct weights by themselves.

3.3 Case Study: Anesthetist Scheduling

In this section, we show how the proposed model can be used to schedule anesthetists over a 2-week planning horizon (see Sect. 3.3.2), whereas in Sect. 3.3.1 the scheduling procedure currently employed at MRI is presented. The example is based on the basic parameter settings that will be introduced in Sect. 5.1.2 and on the demand profile presented in Fig. 5.3. We use the basic model (1)–(14) with some enhancements. The model presented in Sect. 3.2.3 and given by (7b, 8b, 27–31) will be used for the break placements. Furthermore, we apply time window constraints (see Sect. 3.2.2) and on-call service constraints (see Sect. 3.2.1). In this example we have neither vacation periods for some anesthetists nor holiday periods.

3.3.1 Current Practice

Since one of our goals is to compare the rosters generated by our model with current practice, it is necessary to outline the scheduling procedures now in use. At MRI, all anesthetists start at 7:30 a.m. each weekday morning and work a fixed shift of 9 h, which includes a 30-min break. To deal with uncovered demand after the shifts end at 4:30 p.m., schedulers follow an unofficial rule that requires a specified number of physicians, for example 2 in a large operating theater (like the central operating theater), to stay beyond their regular shift. This additional time is paid out as overtime. This rule is really a stopgap measure because it often leads to a violation of the general labor agreement clause that limits a workday to 12 h (see [88]). Furthermore, overtime is needed when operations unexpectedly last longer than 4:30 p.m. All overtime is paid out, thus a compensation of overtime is not considered in current practice.

Following these procedures, the best schedule that could be generated manually required 161 h of paid overtime. At a conservative wage rate of 22.5 Euros per hour, this translates into a cost of 3,622.5 Euros for the central operating theater alone for just two weeks. In addition, because breaks were not explicitly stated in the schedule, it was often impossible for the anesthetists to take them, at least in accordance with the rules specified in their labor contract. Finally, a considerable amount of overtime was needed to cover demand after 4:30 p.m.

Figure 3.5 displays the coverage provided by the schedules generated manually for the 2-week planning horizon in January 2005. The dark area corresponds to the demand (cf. Fig. 5.3), whereas the lighter gray area indicates the cumulative supply.

3.3 Case Study: Anesthetist Scheduling

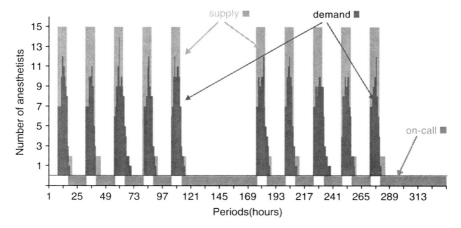

Fig. 3.5 Demand/supply profile corresponding to current practice

The rectangles of height 2 in the supply correspond to the assignment of planned overtime based on the unofficial rule, while the darker gray uniform rectangles at the bottom of the figure correspond to the on-call service. All the light areas, except for the on-call service, indicate that there are anesthetists on duty but that they are idle; however, all demand after 4:30 p.m. has to be covered with overtime.

3.3.2 Solution of the Model

To generate the following solution we have used the heuristic decomposition strategy which will be presented in Sect. 4.2.

Figure 3.6 displays the schedule for anesthetists 1–8 over the first week. The complete schedule over the 2-week planning horizon generated by our model is provided in Fig. A.3 in the Appendix. Recall that a sequence of 1s indicates a shift, a 0 in the respective shift denotes a break assignment, and the remaining 0s in the column indicate that the anesthetist is off duty. The lighter gray shading signals a regular shift while the darker gray shading signals an on-call service. As can be seen, during the week we have both on-call services and regular shifts, whereas on the weekends, only on-call service is scheduled – one 24-h shift each day (see Sect. 3.2.1).

In what follows we detail the solution provided by our model for 16 anesthetists. Final schedules produced by the model do not call for any overtime despite the assignment of hour-long breaks to all shifts, and in many cases, provide a significant amount of undertime which is displayed in Table 3.1. In total, final rosters consist of 62 h of undertime in the first week and 92 h of undertime in the second week. None of the undertime goes to compensate for assigned overtime in the previous week. These statistics suggest that the department is overstaffed. In general, spare

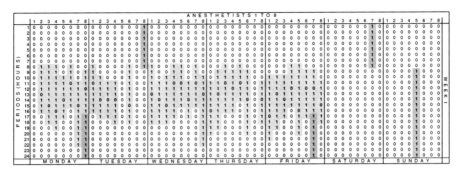

Fig. 3.6 Part of the duty roster produced by model for all anesthetists

Table 3.1 Undertime results for 2-week example

	Anesthetist															
Week	1	2	3	4	5	6	7	8	9	10	11	12	13	14	15	16
1	4	1	3	1	10	0	16	0	9	3	7	2	0	0	5	1
2	9	1	8	2	2	1	9	9	12	16	4	7	1	2	2	7

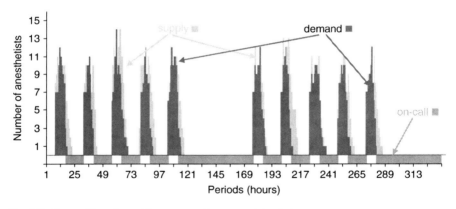

Fig. 3.7 Demand/supply profile corresponding to model solution

time can be used to fill in for absent colleagues, to handle emergencies as they arise or in our model to compensate the remaining hours of an on-call service that are currently paid out (see Sect. 3.2.1).

Figure 3.7 displays the original demand profile (see Fig. 5.3) and the final supply profile generated by the model. The figure shows that on several days the number of anesthetists available at the end of the day exceeds the required number; e.g., on Tuesday there is no demand after 4 p.m. but the lighter gray rectangles indicate that several anesthetists are on duty. This imbalance is a consequence, in part, of the need to assign each anesthetist shifts of minimal length, to consider a proper

3.3 Case Study: Anesthetist Scheduling

Table 3.2 On-call assignments for 2-week example

Week 1							Week 2						
Mo	Tu	We	Th	Fr	Sa	Su	Mo	Tu	We	Th	Fr	Sa	Su
8	15	9	10	7	11	5	15	12	7	13	8	11	9
						Anesthetist							

break assignment and to fulfill the starting time window conditions in the rosters. It is common in the service industry to have schedules with idletime and overtime due to a partial mismatch between supply and demand. The idletime is the difference between the on duty time for all anesthetists and the total demand. In the example, the cumulative demand is 455 h for week one and 447 h for two, while the total number of anesthetist hours available is 672, assuming 16 anesthetists each working 42 h regular per week, considerably more than required. The surplus is characteristic for real world applications especially in hospitals where demand is far from uniform. Nevertheless, by allowing some flexibility in break assignments and starting times, it is possible to dramatically improve current practice with respect to cost and the effective use of staff.

The on-call assignments for the 2-week planning horizon are given in Table 3.2 as well as in Fig. A.3 in the Appendix. The results indicate that several anesthetists have two on-call services in their complete rosters, e.g. 8 and 15, while the others have either zero or one. For MRI, this is a desirable feature of the solution because some physicians prefer the extra pay that comes with the on-call service. Of course, it is an easy matter to limit the number of on-call services that can be assigned to a physician by bounding the sum of the $y_{j,i,t}^{oc}$ variables in the complete planning horizon.

Chapter 4
Solution Methodologies

In this chapter we present solutions methodologies for the flexible shift scheduling problem of physicians in a hospital. After discussing some preprocessing issues in Sect. 4.1 we present a heuristic decomposition strategy in Sect. 4.2 which has been applied to find a set of rosters in Sect. 3.3. Finally, we conclude the chapter by presenting our new column generation approach embedded in a B&B framework to find optimal solutions to the physician scheduling problem. This combination is widely known under B&P algorithms.

4.1 Preprocessing

The number of variables and constraints in the model presented in Chap. 3 with all enhancements is approximately $O(\sum_{j \in \mathcal{J}} |\mathcal{I}_j| \cdot (2 \cdot |\mathcal{T}| + |\mathcal{L}|))$, which suggests a formidable computational burden for even modest size instances. To reduce the number of variables, we exploit the inherent structure of the demand profile by taking into account minimum and maximum shift lengths and the need for a minimum number of rest periods between shifts. Taking advantage of the problem specific data we define ranges for shift starts and rest starts for each group of physicians j and hence reduce the number of variables.

In a preprocessing step we first determine the earliest and latest periods in a day that have demand. Given these values, we then determine the earliest and latest periods over the week that should be considered for starting a shift on any day where the earliest necessary start over the week is calculated by taking the minimum of all earliest demand periods (first time where demand occurs on a day within the week considered). In contrast to calculate the latest shift starting, we determine the latest demand period in a week by taking the maximum of all latest demand periods on various days within the week and then subtract the minimum shift length to determine the latest starting period for a shift. For instance, if the latest demand occurrence in an arbitrary week is between 8 and 9 p.m. which corresponds to period 21 within a day and we have to assign shifts with at least six periods, then it is sufficient to start the latest shift that covers the demand in period 21 (8–9 p.m.) at the beginning of period 16 corresponding to starting the shift at 3 p.m. on that

J.O. Brunner, *Flexible Shift Planning in the Service Industry*, Lecture Notes in Economics and Mathematical Systems 640, DOI 10.1007/978-3-642-10517-3_4, © Springer-Verlag Berlin Heidelberg 2010

39

day. In conclusion, if avoidable, there is no reason to cover periods in which there is no demand. These calculations allow us to limit the number of $y_{j,i,t}^{shift}$ variables and to fix corresponding $x_{j,i,t}$ and $x_{j,i,t}^{break}$ variables in the model. Recall, when preprocessing is applied each group of physicians $j \in \mathcal{J}$ is treated separately.

The same idea of limiting the periods in which a rest period can start applies when one takes into account the fact that a shift must have a minimum length. In particular, when a shift starts in the first period of the possible start time range on a day, the earliest it can end is $\underline{T}_j^{shift} - 1$ periods later. This allows us to limit the periods in which the $y_{j,i,t}^{rest}$ variables are defined. On the weekends in case no demand occurs ($d_t = 0$), of course, all $x_{j,i,t}$ and $x_{j,i,t}^{break}$ can be fixed; all $y_{j,i,t}^{shift}$ and $y_{j,i,t}^{rest}$ can be omitted.

For those scenarios in which starting time ranges are defined, we introduce special order set (SOS) of type 1 constraints (cf. [121, 123]) for the $y_{j,i,t}^{shift}$ variables for each day. SOS are sets that include a set of decision variables. The type 1 description says that at most one variable in the SOS can be greater than zero. In comparison to that a SOS type 2 constraint consists of a set of decision variables where two adjacent variables in the set can be greater than zero. We will present an example for SOS type 1 constraints shortly. Let the set $\mathcal{T}_{j,w}^{earlystart}$ contains all earliest starting periods within the week w for physician group j and the parameter $T_{j,w}^{startlength}$ denotes the length of the starting time range which is the latest shift start minus the earliest shift start. The following constraints are introduced as SOS constraints.

$$\sum_{\tau=t}^{t+T_{j,w}^{startlength}} y_{j,i,\tau}^{shift} \leq 1, \forall j \in \mathcal{J}, i \in \mathcal{I}_j, w \in \mathcal{W}, t \in \mathcal{T}_{j,w}^{earlystart} \tag{41}$$

For example, the earliest shift start within a week is at 7 a.m. (period 8 on any day) and the latest shift start is at 3 p.m. (period 16). Furthermore, we assume five normal working days, Monday through Friday, where regular shifts are assigned and 24 periods within a day (i.e. $T^{day} = 24$), then for an arbitrary group j we have $\mathcal{T}_{j,w}^{earlystart} = \{t \in \mathcal{T}_w : t \bmod 24 = 8 \cap t < 5 \cdot 24\} = \{8, 32, 56, 80, 104\}$ and $T_{j,w}^{startlength} = 8 \ (= 16 - 8)$ and the first SOS type 1 constraint for physician i associated with group j is

$$\sum_{\tau=8}^{16} y_{j,i,\tau}^{shift} \leq 1 \tag{42}$$

To construct the SOS type 1 constraints according to the previous description is possible in our case because the demand data imply that all demand on any day can be covered by shifts that start on that day. For the more general case, the planning horizon is divided into intervals of length equal to $\underline{T}_j^{shift} + \underline{T}_j^{rest}$ in order to construct the SOS type 1 constraints (41).

In addition, it is necessary to consider the implications of an on-call service on the demand profile. Recall that physicians who are assigned an on-call service on a normal working day, work the first 8 h as a regular (fixed) shift beginning at 8 a.m. (period 9, 33, 57 and so forth) on the specified day (cf. Sect. 3.2.1). From that it follows that we must reduce the demand profile for periods 9 through 17 by d_t^{oc} on each normal working day corresponding to $t \in \mathcal{L}$. Note that d_t^{oc} can have various forms for different t but in our case we assume it to be equal for all $t \in \mathcal{L}$ and hence we might omit the index t. Because a pure on-call service – 16 h on normal working days and 24 h on Saturday and Sunday – does not count towards demand coverage, we do not reduce the demand profile for that time. If a break is to be assigned to the regular (fixed) shift, then the demand during the break period must be increased by the number of physicians, d_t^{oc}, who are on-call that day. For all the computations, we assume that the break for the fixed shift before an on-call service on a normal working day always occurs between 11 and 12 a.m., i.e. noon, on each regular working day.

4.2 Heuristic Decomposition Strategy

To reduce the computational burden, we now propose a decomposition heuristic in which a $|\mathcal{W}|$ week problem is broken down into $|\mathcal{W}|$ 1-week problems. Here we concentrate on 1-week subproblems, however the strategy can also be embodied to any multi week decomposition of the original planning horizon. In this approach, the results obtained by solving the basic problem for a predecessor week serves as input for the following week. We call this a *rolling time horizon in* previous statements. What is critical is keeping track of the number of hours worked, especially overtime in the first week, because it can be compensated for by scheduling an equal amount of undertime in the second week. This logic carries forward for any pair of adjacent weeks (recall that it is not possible to compensate for undertime in 1 week by assigning an equal amount of overtime in the next). It is also necessary to keep track of the on-call assignments and, in the more general case where regular shifts are assigned on the weekend, to keep track of the Sunday shifts (to count the rest periods between shifts, it is necessary to know the last working period on Sunday). The following additional parameters are needed to define the 1-week model.

Parameters

$o_{j,i}^{prev}$	Initial overtime from the previous week for each $i \in \mathcal{I}_j$ associated with group $j \in \mathcal{J}$
$s_{j,i}^{prev}$	1, if physician $i \in \mathcal{I}_j$ associated with group $j \in \mathcal{J}$ is assigned an on-call service starting on the last day of the previous week, 0 otherwise
T_{spill}^{oc}	Number of hours that an on-call service spills over to the following week; $T_{spill}^{oc} = 8$ in our case

Now, discarding the weekly index w, the updated model remains the same except for the constraints (9) which are replaced by

$$h_{j,i} \geq o_{j,i}^{prev} - u_{j,i}, \forall j \in \mathcal{J}, i \in \mathcal{I}_j \tag{9a}$$

and two new sets of constraints. The first assures the minimum rest for the anesthetist who is on-call on Sunday in the previous week and the second forbids that if he is assigned to an on-call service on Sunday, then he will not be assigned to one on Monday.

$$\sum_{t=1}^{\underline{T}_j^{rest}+T_{spill}^{oc}-1} (1 - x_{j,i,t}) \geq (\underline{T}_j^{rest} + T_{spill}^{oncall}) \cdot s_{j,i}^{prev}, \forall j \in \mathcal{J}, i \in \mathcal{I}_j \tag{43}$$

$$y_{j,i,t}^{oc} + s_{j,i}^{prev} \leq 1, \forall j \in \mathcal{J}, i \in \mathcal{I}_j, t \in \{l \in L : l < T^{day}\} \tag{44}$$

To simplify the presentation, it is assumed that $o_{j,i}^{prev} = 0$, $\forall j \in \mathcal{J}, i \in \mathcal{I}_j$, in (9a) for the initial week of the planning problem. This assumption is made for all the computations in Chap. 5.

As will be seen in Sect. 5.2 the decomposition heuristic outperforms the full 2-week MIP with respect to the objective function value. Furthermore, its runtime increases only linearly with the number of weeks, which is an important consideration when solving real-world instances over the year. However, the decomposition is just an heuristic as the following example shows.

We assume that there are no break assignments, time window constraints, and on-call services to facilitate the presentation. The planning horizon is 2 weeks but just Monday through Friday are normal working days. Furthermore, we consider two identical physicians with 42 regular working hours per week. We have a uniform demand profile that can be covered completely with regular staff by assumption. There are two physicians needed on Monday from period eight to period 17, on Tuesday through Friday from period eight to period 15 (see Table 4.1). From that it follows, that each of the physicians has to work 42 h (a 10-h shift on Monday and an 8-h shift on Tuesday–Friday). Additionally, there is a request for one physician on Thursday and on Friday in period 16. Table 4.1 shows the demand and the rosters for both physicians in the first week. In detail, the first column shows the period t. The next three columns are the demand d_t, the assignment for physician 1 and the assignment for physician 2 on Monday. The remaining columns give the demand and assignments for Tuesday through Friday.

When we solve the problem for the first week with the decomposition strategy two possible (optimal) solutions could arise. Indeed there are four optimal solutions when symmetry is considered. The solutions are:

1. Physician 1 works 42 h and physician 2 works 44 h. Hereby, physician 2 is on duty on Thursday and on Friday in period 16. That solution is identical to the solution when physician 1 works 44 and physician 2 works 42 h (symmetry).

4.2 Heuristic Decomposition Strategy

Table 4.1 First week solution

| t | d_t | Phys | | d_t | Phys | | d_t | Phys | | d_t | Phys | | d_t | Phys | |
		1	2		1	2		1	2		1	2		1	2
1	0	0	0	0	0	0	0	0	0	0	0	0	0	0	0
2	0	0	0	0	0	0	0	0	0	0	0	0	0	0	0
3	0	0	0	0	0	0	0	0	0	0	0	0	0	0	0
4	0	0	0	0	0	0	0	0	0	0	0	0	0	0	0
5	0	0	0	0	0	0	0	0	0	0	0	0	0	0	0
6	0	0	0	0	0	0	0	0	0	0	0	0	0	0	0
7	0	0	0	0	0	0	0	0	0	0	0	0	0	0	0
8	2	1	1	2	1	1	2	1	1	2	1	1	2	1	1
9	2	1	1	2	1	1	2	1	1	2	1	1	2	1	1
10	2	1	1	2	1	1	2	1	1	2	1	1	2	1	1
11	2	1	1	2	1	1	2	1	1	2	1	1	2	1	1
12	2	1	1	2	1	1	2	1	1	2	1	1	2	1	1
13	2	1	1	2	1	1	2	1	1	2	1	1	2	1	1
14	2	1	1	2	1	1	2	1	1	2	1	1	2	1	1
15	2	1	1	2	1	1	2	1	1	2	1	1	2	1	1
16	2	1	1	0	0	0	0	0	0	1	?	?	1	?	?
17	2	1	1	0	0	0	0	0	0	0	0	0	0	0	0
18	0	0	0	0	0	0	0	0	0	0	0	0	0	0	0
19	0	0	0	0	0	0	0	0	0	0	0	0	0	0	0
20	0	0	0	0	0	0	0	0	0	0	0	0	0	0	0
21	0	0	0	0	0	0	0	0	0	0	0	0	0	0	0
22	0	0	0	0	0	0	0	0	0	0	0	0	0	0	0
23	0	0	0	0	0	0	0	0	0	0	0	0	0	0	0
24	0	0	0	0	0	0	0	0	0	0	0	0	0	0	0
	Mon			Tue			Wed			Thu			Fri		

2. Both physicians work 43 h. Each of them covers one of the periods 16 on Thursday or on Friday (consider also symmetry).

Given the following demand profile for the second week (see Table 4.2), we argue that the decomposition is not an exact method. There are two physicians requested on Monday–Friday from period eight to period 15. Additionally, one physician is called in on Friday in period 16 and 17. Then one physician works 42 h and the other works 40 h. There is no solution where both work 41 h and the demand is covered completely. For that case, if the decomposition yields the first solution given above we obtain an optimal solution for the two-week planning horizon with the decomposition, since all the overtime needed in the first week could be compensated by undertime in the second. However, if the decomposition yields the second solution due to the demand profile in the second week only 1 h of overtime could be compensated by undertime. Consequently, the other hour of overtime has to be paid out and the decomposition gives a non-optimal solution to the full 2-week model. Therefore, it would be better to assign both hours of overtime to one physician in the first week as assigning 1 h of overtime to each of them given the demand in the

Table 4.2 Second week solution alternative 1

| t | d_t | Phys | | d_t | Phys | | d_t | Phys | | d_t | Phys | | d_t | Phys | |
		1	2		1	2		1	2		1	2		1	2
1	0	0	0	0	0	0	0	0	0	0	0	0	0	0	0
2	0	0	0	0	0	0	0	0	0	0	0	0	0	0	0
3	0	0	0	0	0	0	0	0	0	0	0	0	0	0	0
4	0	0	0	0	0	0	0	0	0	0	0	0	0	0	0
5	0	0	0	0	0	0	0	0	0	0	0	0	0	0	0
6	0	0	0	0	0	0	0	0	0	0	0	0	0	0	0
7	0	0	0	0	0	0	0	0	0	0	0	0	0	0	0
8	2	1	1	2	1	1	2	1	1	2	1	1	2	1	1
9	2	1	1	2	1	1	2	1	1	2	1	1	2	1	1
10	2	1	1	2	1	1	2	1	1	2	1	1	2	1	1
11	2	1	1	2	1	1	2	1	1	2	1	1	2	1	1
12	2	1	1	2	1	1	2	1	1	2	1	1	2	1	1
13	2	1	1	2	1	1	2	1	1	2	1	1	2	1	1
14	2	1	1	2	1	1	2	1	1	2	1	1	2	1	1
15	2	1	1	2	1	1	2	1	1	2	1	1	2	1	1
16	0	0	0	0	0	0	0	0	0	0	0	0	1	?	?
17	0	0	0	0	0	0	0	0	0	0	0	0	1	?	?
18	0	0	0	0	0	0	0	0	0	0	0	0	0	0	0
19	0	0	0	0	0	0	0	0	0	0	0	0	0	0	0
20	0	0	0	0	0	0	0	0	0	0	0	0	0	0	0
21	0	0	0	0	0	0	0	0	0	0	0	0	0	0	0
22	0	0	0	0	0	0	0	0	0	0	0	0	0	0	0
23	0	0	0	0	0	0	0	0	0	0	0	0	0	0	0
24	0	0	0	0	0	0	0	0	0	0	0	0	0	0	0
	Mon			Tue			Wed			Thu			Fri		

second week. Nevertheless, we cannot guarantee that the decomposition gives us the first solution.

In the sequential, we consider a case where it is better to get the second solution in the first week. If there are two physicians requested on Monday through Thursday from period eight to period 15 and on Friday from period eight to period 16 (cf. Table 4.3) then it would be better to assign both of them 1 h of overtime in the first week. In that case, each physician has to work 41 h in the second week (only optimal solution) and both overtime hours form the first week could be compensated.

We have proven that the decomposition does not guarantee optimal solutions since an optimal decision in the future depends on current decisions.

4.3 Column Generation and B&P Algorithm

In this section we present our new column generation approach (CGA) to solve the flexible shift scheduling problem of physicians at hospitals. We start with stating the master problem and the subproblem(s) in Sects. 4.3.1 and 4.3.2, respectively.

4.3 Column Generation and B&P Algorithm

Table 4.3 Second week solution alternative 2

		Phys			Phys			Phys			Phys			Phys	
t	d_t	1	2	d_t	1	2	d_t	1	2	d_t	1	2	d_t	1	2
1	0	0	0	0	0	0	0	0	0	0	0	0	0	0	0
2	0	0	0	0	0	0	0	0	0	0	0	0	0	0	0
3	0	0	0	0	0	0	0	0	0	0	0	0	0	0	0
4	0	0	0	0	0	0	0	0	0	0	0	0	0	0	0
5	0	0	0	0	0	0	0	0	0	0	0	0	0	0	0
6	0	0	0	0	0	0	0	0	0	0	0	0	0	0	0
7	0	0	0	0	0	0	0	0	0	0	0	0	0	0	0
8	2	1	1	2	1	1	2	1	1	2	1	1	2	1	1
9	2	1	1	2	1	1	2	1	1	2	1	1	2	1	1
10	2	1	1	2	1	1	2	1	1	2	1	1	2	1	1
11	2	1	1	2	1	1	2	1	1	2	1	1	2	1	1
12	2	1	1	2	1	1	2	1	1	2	1	1	2	1	1
13	2	1	1	2	1	1	2	1	1	2	1	1	2	1	1
14	2	1	1	2	1	1	2	1	1	2	1	1	2	1	1
15	2	1	1	2	1	1	2	1	1	2	1	1	2	1	1
16	0	0	0	0	0	0	0	0	0	0	0	0	2	1	1
17	0	0	0	0	0	0	0	0	0	0	0	0	0	0	0
18	0	0	0	0	0	0	0	0	0	0	0	0	0	0	0
19	0	0	0	0	0	0	0	0	0	0	0	0	0	0	0
20	0	0	0	0	0	0	0	0	0	0	0	0	0	0	0
21	0	0	0	0	0	0	0	0	0	0	0	0	0	0	0
22	0	0	0	0	0	0	0	0	0	0	0	0	0	0	0
23	0	0	0	0	0	0	0	0	0	0	0	0	0	0	0
24	0	0	0	0	0	0	0	0	0	0	0	0	0	0	0
	Mon			Tue			Wed			Thu			Fri		

Then, after giving an overview of the CGA-technique we introduce two branching strategies in Sects. 4.3.4 and 4.3.5. We show the validity of the proposed strategies by looking at the dual of MP in Sect. 4.3.6. Heuristics to find an initial set of columns and feasibility heuristics are presented in Sect. 4.3.7. Eventually, we conclude with some new ideas to improve the performance of the algorithm (see Sect. 4.3.8).

Generally speaking, column generation is an iterative procedure that considers only a subset of feasible columns (schedules) at a time and generates new columns via one or more separated optimization problem(s), the so-called subproblem(s), on an as needed basis (e.g., see [57, 122, 123]). The coordination of the columns is done in the master problem. The procedure can be used as an alternative to the simplex algorithm to solve an LP or the linear relaxation(s) of a general IP within a B&B framework, the focus of the following research, and is successful to solve large-scale problems.

In general, shift and related scheduling problems can be modeled in two distinct but equivalent ways. The first uses a set covering formulation in which each column corresponds to a feasible schedule. This is called the extensive form (e.g., see [57]). The postulate in set covering problems is that each element in the ground set must be covered at least once. In our case, the ground set is the demand. The major drawback of this construct is the vast number of feasible schedules (columns) that have to be considered. The second formulation takes a shift-based view in which all system and individual constraints are explicitly written as rows in the model. This is called the original or compact form (e.g., see [57]).

In Chap. 3 we proposed an implicit shift-based model for the flexible shift scheduling problem of physicians in a hospital. Furthermore, we have presented a heuristic decomposition strategy in the previous section. As will be seen soon in Sect. 5.2 the results obtained by the heuristic decomposition indicate that the strategy worked well and that the 1-week subproblems are relatively easy to solve with CPLEX. However, when we use a 2-week decomposition, the corresponding subproblems required too much computational effort, limiting our ability to prove optimality for planning horizons beyond one week when subproblems were solved. The primary reason for this situation is the lack of improvement in the lower bound obtained at the root node of the search tree set up by CPLEX. In an attempt to overcome this hurdle and extend the planning horizon of solvable instances, we develop a column generation scheme that is designed to provide improved lower bounds and overall better performance. In theory, when using a set covering formulation and generating columns iteratively by solving IP subproblem(s), the lower bound should be tighter than that obtained by solving the shift-based formulation as a linear program (cf. [123]). If the subproblems are not IPs then the column generation algorithm will provide the same bound obtained by the linear programming relaxation of the original IP (see [72]).

4.3.1 Master Problem Formulation

To implement CGA a *master problem* (MP) is created using a set covering-type formulation. To do this, we separate the regular and on-call demand constraints, which cut across all physicians in the original compact formulation, from the remaining constraints, which decompose by physician group. The latter constitute the feasible region of what are called the *pricing subproblems* (SPs). This reformulation reduces the original integer program to one of smaller row dimension, where each column corresponds to an extreme point of the separated physician group constraints (because the accompanying feasible region is bounded, there are no extreme rays). When solving the LP relaxation of MP, say, during B&B, a brute force enumeration scheme would yield an optimal solution when MP contains all feasible columns. The impracticality of such a scheme, though, is well known so a restricted version of MP is usually solved starting with a small subset of columns and then solves a series of subproblems to identify promising additions. Like in the simplex method, solving one or more subproblem determines the entering variable(s) and proves optimality by implicitly pricing out all absent columns.

When all the physicians are identical, we obtain one SP that serves as the pattern or column generator for MP. Our model considers the more general case in which there is an SP for each physician group assuming that the physicians within a

4.3 Column Generation and B&P Algorithm

group are interchangeable (independent of i). This means that exchanging the final assignments for two physicians within the same group leads to an essential identical solution. The aggregation of all physicians within a group to one SP as pattern generator eliminates symmetry in the solution space but original binary variables that determine the assignment are replaced by general integer variables. To state MP as a set covering-type model the following additional notation is used.

Indices and sets

k	Index for schedules
$\mathcal{K}(j)$	Set of schedules for physician type $j \in \mathcal{J}$
$\mathcal{K}$	Set of all schedules with $\mathcal{K} = \bigcup_{j \in \mathcal{J}} \mathcal{K}(j)$

Parameters

c_{jk}^{sched}	Cost associated with schedule (column) k for physician type j
n_j^{phy}	Total number of type j physicians that are available
$X_t^{j,k}$	1, if schedule k for type j covers regular demand in period $t \in T^{dem}$, 0 otherwise
$Y_t^{j,k}$	1, if schedule k for type j has a break in period $t \in T^{dem}$, 0 otherwise
$Z_t^{j,k}$	1, if schedule k for type j covers an on-call service starting in period $t \in \mathcal{L}$, 0 otherwise

General integer decision variables

$\lambda_{j,k}$	Number of physicians of type j who are assigned to schedule k

Master problem

$$\text{Min} \sum_{j \in \mathcal{J}} \sum_{k \in \mathcal{K}(j)} c_{j,k}^{sched} \cdot \lambda_{j,k} + \sum_{t \in T^{dem}} c^{out} \cdot x_t^{out} + \sum_{t \in \mathcal{L}} c^{oc\_out} \cdot x_t^{oc\_out} \qquad (45)$$

subject to

$$\sum_{j \in \mathcal{J}} \sum_{k \in \mathcal{K}(j)} \left(X_t^{j,k} - Y_t^{j,k} \right) \cdot \lambda_{j,k} + x_t^{out} \geq d_t, \forall t \in T^{dem} \qquad (46)$$

$$\sum_{j \in \mathcal{J}} \sum_{k \in \mathcal{K}(j)} Z_t^{j,k} \cdot \lambda_{j,k} + x_t^{oc\_out} \geq d_t^{oc}, \forall t \in \mathcal{L} \qquad (47)$$

$$\sum_{k \in \mathcal{K}(j)} \lambda_{j,k} \leq n_j^{phy}, \forall j \in \mathcal{J} \tag{48}$$

$$\lambda_{j,k}, \ x_t^{out}, \ x_l^{oc\_out} \geq 0 \text{ and integer}, \forall j \in \mathcal{J}, k \in \mathcal{K}(j), t \in \mathcal{T}^{dem}, l \in \mathcal{L} \tag{49}$$

The objective function (45) minimizes the cost of the assigned physicians to schedules over all types j and the cost incurred when there are gaps in coverage. In the first term, the cost of a roster (column) depends on the physician type j and is defined as

$$c_{j,k}^{sched} = \sum_{w \in \mathcal{W}} \left(c_j^{paid} \cdot h_{j,w} + c_j^{over} \cdot o_{j,w} \right), \forall j \in \mathcal{J}, k \in \mathcal{K}(j) \tag{50}$$

where c_j^{paid} is the cost per paid out hour and c_j^{over} is the cost per hour of overtime such that $c_j^{over} > c_j^{paid}, \forall j \in \mathcal{J}$. Because $h_{j,w}$ and $o_{j,w}$ are not present in MP but in SP_j, the cost coefficients $c_{j,k}^{sched}$ have to be calculated independently for each j and k. The variable $h_{j,w}$ denotes the paid out time in week $w \in \mathcal{W}$ which is equal to the overtime from the previous week minus the undertime in the current week; the variable $o_{j,w}$ denotes the assigned overtime in a week w for physician type j. Both $h_{j,w}$ and $o_{j,w}$ are determined from the solution of SP_j stated presently.

Constraints (46) ensure that the total number of physicians that are on duty to cover the regular demand in each period $t \in \mathcal{T}^{dem}$ is sufficient. The general integer variable $\lambda_{j,k}$ corresponds to the number of physicians of type j who are assigned to schedule $k \in \mathcal{K}(j)$ and the variable x_t^{out} is used to guarantee feasibility of MP. As mentioned previously, this variable signals the option to hire outside physicians to cover the requests in period t. Constraints (47) force the number of physicians who are given an on-call service starting in period $t \in \mathcal{L}$ to be at least d_t^{oc}. Like x_t^{out}, the variable $x_t^{oc\_out}$ serves to guarantee feasibility. Recognizing that shifts span a fixed number of time periods, both (46) and (47) are written as inequalities rather than equalities for practical purposes. When more than d_t^{oc} physicians are assigned to period $t \in \mathcal{L}$ a simple post-processing procedure is applied to delete excess coverage and to update the effected rosters accordingly.

The next set of constraints (48) limits the assignment of physicians in each group j to no more than the number of physicians available n_j^{phy}. This is achieved by summing the assignment variables $\lambda_{j,k}$ over all $k \in \mathcal{K}(j)$ for each $j \in \mathcal{J}$. Variable definitions are given in (49); however, the integrality conditions on x_t^{out} and $x_t^{oc_{o}ut}$ can be relaxed because they will always be integer when $\lambda_{j,k}$ are integer. This is a direct result of the fact that we are minimizing a linear function (45) of these variables and all the data in the problem, $X_t^{j,k}$, $Y_t^{j,k}$, $Z_l^{j,k}$, d_t and d_l^{oc} are integral for all $t \in \mathcal{T}^{dem}$ and $l \in \mathcal{L}$.

4.3.2 Subproblem Formulation

As mentioned, it is impractical to include all possible columns in MP so we work only with a small subset, which defines the *restricted master problem* (RMP). Following the standard approach, CGA is used to solve the LP relaxation of RMP obtained by treating the $\lambda_{j,k}$ variables as continuous in (45)–(49). At the termination, however, if all $\lambda_{j,k}$ are integral, then we have solved the original problem and present the rosters in the optimal solution as the final assignments. Otherwise, B&B is performed with CGA being used to find lower bounds at each node in the search tree.

To derive the objective function of the pricing subproblems used to generate new columns for MP, we need the duals associated with an optimal solution to RMP. Let $\delta_t^{dem} \geq 0$ and $\delta_t^{oc} \geq 0$ be the dual variables of the regular demand constraints (46) and the on-call demand constraints (47), respectively, for time period t, and let $\delta_j^{conv} \leq 0$ be the dual variable of the convexity constraint (48) associated with physician type j. With that in mind we can define a generic reduced cost of a column in RMP as follows.

$$\overline{c}_{j,k} = c_{j,k}^{sched} - \sum_{t \in \mathcal{T}^{dem}} \delta_t^{dem} \cdot \left(X_t^{j,k} - Y_t^{j,k} \right) - \sum_{t \in \mathcal{L}} \delta_t^{oc} \cdot Z_t^{j,k} - \delta_j^{conv} \quad (51)$$

For situations in which physicians in some group j, such as part-timers, do not work on-call services, then the assignment parameters $Z_t^{j,k} = 0$ for all $t \in \mathcal{L}$. Hence, the corresponding dual variables δ_t^{oc} would be omitted from (51) when calculating the reduced cost of a roster $k \in \mathcal{K}(j)$.

Dual feasibility is achieved when all columns in RMP have nonnegative reduced cost and it is not possible to identify any absent columns whose reduced cost would be negative given the current values of $\{\delta_t^{dem}, \delta_l^{oc}, \delta_j^{conv} : t \in \mathcal{T}^{dem}, l \in \mathcal{L}, j \in \mathcal{J}\}$.

To verify that no columns exist such that $\overline{c}_{j,k} < 0$ for all $j \in \mathcal{J}, k \in \mathcal{K}(j)$, we minimize $\overline{c}_{j,k}$ over the constraint region associated with physician group j. Dropping the index k in (51) and replacing all parameters with their variable counterparts leads to

$$\overline{c}_j = \sum_{w \in \mathcal{W}} \left(c_j^{paid} \cdot h_{j,w} + c_j^{over} \cdot o_{j,w} \right)$$
$$- \sum_{t \in \mathcal{T}^{dem}} \delta_t^{dem} \cdot \left(x_{j,t} - y_{j,t}^{break} \right) - \sum_{t \in \mathcal{L}} \delta_t^{oc} \cdot y_{j,t}^{oc} - \delta_j^{conv} \quad (51a)$$

which defines the objective function for SP_j. The following modified (we drop the index i) and additional notation is used in the developments.

50   4 Solution Methodologies

Parameters

δ_t^{dem}	Dual value of MP regular demand constraint set (46) for $t \in \mathcal{T}^{dem}$
δ_t^{oc}	Dual value of MP on-call demand constraint set (47) for $t \in \mathcal{L}$
δ_t^{dem}	Dual value of MP convexity constraint set (48) for $j \in \mathcal{J}$

Parameters for physician group j

c_j^{paid}	Cost per hour of paid out time
c_j^{over}	Cost per hour of overtime
$\overline{o}_j$	Maximal allowed overtime in a week
r_j	Regular working hours per week according to the labor contract

Binary decision variables for physician group j

$x_{j,t}$	1, if on duty in period t, 0 otherwise
$y_{j,t}^{shift}$	1, if a shift begins in period t, 0 otherwise
$y_{j,t}^{rest}$	1, if rest period in period t, 0 otherwise
$y_{j,t}^{oc}$	1, if an on-call service begins in period $t \in \mathcal{L}$, 0 otherwise

General integer decision variables for physician group j

$o_{j,w}$	Amount of overtime in week w
$u_{j,w}$	Amount of undertime in week w
$h_{j,w}$	Amount of paid out time in week w
$e_{j,w}$	Earliest starting time for a shift in week w
$l_{j,w}$	Latest starting time for a shift in week w

Subproblem j (SP_j)

$$
\begin{aligned}
\text{Min} \sum_{w \in \mathcal{W}} &\left(c_j^{paid} \cdot h_{j,w} + c_j^{over} \cdot o_{j,w} \right) \\
&- \sum_{t \in \mathcal{T}^{dem}} \delta_t^{dem} \cdot \left(x_{j,t} - y_{j,t}^{break} \right) - \sum_{t \in \mathcal{L}} \delta_t^{oc} \cdot y_{j,t}^{oc} - \delta_j^{conv}
\end{aligned}
\tag{52}
$$

4.3 Column Generation and B&P Algorithm

subject to

$$y_{j,t}^{shift} = x_{j,t} \cdot (1 - x_{j,t-1}), \forall t \in \mathcal{T} \tag{53}$$

$$y_{j,t}^{rest} = x_{j,t-1} \cdot (1 - x_{j,t}), \forall t \in \mathcal{T} \tag{54}$$

$$\sum_{\tau=t}^{\min(t+\underline{T}_j^{shift}-1,|\mathcal{T}|)} x_{j,\tau} \geq \min\left(\underline{T}_j^{shift}, |\mathcal{T}| - t + 1\right) \cdot y_{j,t}^{shift}, \forall t \in \mathcal{T} \tag{55}$$

$$\sum_{\tau=t+\underline{T}_j^{shift}}^{t+\overline{T}_j^{shift}} (1 - x_{j,\tau}) \geq y_{j,t}^{shift}, \forall t \in \left\{1, \ldots, |\mathcal{T}| - \overline{T}_j^{shift}\right\} \tag{56}$$

$$\sum_{\tau=t}^{\min(t+\underline{T}_j^{rest}-1,|\mathcal{T}|)} (1 - x_{j,\tau}) \geq \min\left(\underline{T}_j^{rest}, |\mathcal{T}| - t + 1\right) \cdot y_{j,t}^{rest}, \forall t \in \mathcal{T} \tag{57}$$

$$\sum_{t \in \mathcal{T}_w} x_{j,t} + \sum_{t \in \{l \in \mathcal{L}_w : l > T^{day}\}} f_1(t - T^{day}) \cdot y_{j,t-T^{day}}^{oc}$$
$$= r_j + o_{j,w} - u_{j,w}, \forall w \in \mathcal{W} \tag{58}$$

$$h_{j,w} \geq o_{j,w-1} - u_{j,w}, \forall w \in \mathcal{W} \tag{59}$$

$$\sum_{\tau=\max\left(t-\underline{T}_j^{rest},0\right)}^{t-1} \left(1 - x_{j,\tau}\right) \geq \min\left(\underline{T}_j^{rest}, t - 1\right) \cdot y_{j,t}^{oc}, \forall t \in \mathcal{L} \tag{60}$$

$$\sum_{\tau=t}^{\min\left(t+\underline{T}_j^{rest}+T^{oc}-1,|\mathcal{T}|\right)} (1 - x_{j,\tau}) \geq$$
$$\min\left(\underline{T}_j^{rest} + T^{oc}, |\mathcal{T}| - t + 1\right) \cdot y_{j,t}^{oc}, \forall t \in \mathcal{L} \tag{61}$$

$$\sum_{t \in \mathcal{L}_w} y_{i,t}^{oc} \leq \overline{n}_j^{oc}, \forall w \in \mathcal{W} \tag{62}$$

$$y_{j,t}^{oc} + y_{j,t+T^{day}}^{oc} \leq 1, \forall t \in \left\{ l \in \mathcal{L} : l < |\mathcal{T}| - T^{day} \right\} \tag{63}$$

$$e_{j,w} \leq T^{day} - \sum_{\tau=t}^{t+T^{day}-1} (T^{day} - f_2(\tau)) \cdot y_{j,\tau}^{shift}, \forall w \in \mathcal{W}, t \in \mathcal{S}_w \tag{64}$$

$$l_{j,w} \geq \sum_{\tau=t}^{t+T^{day}-1} f_2(\tau) \cdot y_{j,\tau}^{shift}, \forall w \in \mathcal{W}, t \in \mathcal{S}_w \tag{65}$$

$$l_{j,w} - e_{j,w} \leq T_j^{win}, \forall w \in \mathcal{W} \tag{66}$$

$$\sum_{\tau=t+T_j^{pre}}^{\min\{t+\overline{T}_j^{pre},|\mathcal{T}|\}} y_{j,\tau}^{break} \geq y_{i,t}^{shift}, \forall t \in \mathcal{T} \setminus \{|\mathcal{T}| - T_j^{pre} + 1, \ldots, |\mathcal{T}|\} \tag{67}$$

$$\sum_{\tau=t-\overline{T}_j^{shift}+T_j^{pre}}^{t-T_j^{post}-1} y_{j,\tau}^{break} \geq y_{j,t}^{rest}, \forall t \in \mathcal{T} \setminus \left\{ 1, \ldots, \overline{T}_j^{shift} - T_j^{pre} \right\} \tag{68}$$

$$\sum_{t \in \mathcal{T}} y_{j,t}^{shift} = \sum_{t \in \mathcal{T}} y_{j,t}^{break} \tag{69}$$

$$y_{j,t}^{break} \leq x_{j,t}, \forall t \in \mathcal{T} \tag{70}$$

$$\sum_{t \in \mathcal{V}_j} x_{j,t} \leq 0 \tag{71}$$

$$\sum_{t \in \mathcal{H}} x_{j,t} \leq 0 \tag{72}$$

$$x_{j,t} \text{ binary}, \forall t \in \mathcal{T} \cup \{0\} \tag{73}$$

$$y_{j,t}^{shift}, \ y_{j,t}^{rest}, \ y_{j,t}^{break}, \ y_{j,l}^{oc} \text{ binary}, \forall t \in \mathcal{T}, l \in \mathcal{L} \tag{74}$$

$$x_t^{out}, \ h_{i,w} \geq 0 \text{ and integer}, \forall w \in \mathcal{W}, t \in \mathcal{T} \tag{75}$$

4.3 Column Generation and B&P Algorithm 53

$$0 \leq o_{j,w} \leq \overline{o}_j \text{ and integer}, \forall w \in \mathcal{W} \cup \{0\} \tag{76}$$

$$u_{j,w} \geq 0 \text{ and integer}, \forall w \in \mathcal{W} \tag{77}$$

$$l_{j,w}, e_{j,w} \geq 0 \text{ and integer}, \forall w \in \mathcal{W} \tag{78}$$

The objective function (52) minimizes the reduced cost of a feasible roster for a physician in group j. Constraints (53) and (54) link the x-variables with the y-variables, and constraints (55)–(57) define the shifts implicitly. The number of paid out hours is implicitly determined by constraints (58) and (59) which serve to calculate the over- and undertime in each week w.

On-call assignment rules are embedded in constraints (60)–(63), while constraints (64)–(66) ensure that the earliest shift starting time in any week w is no more than T_j^{win} periods before the latest shift starting time for a specific group j. The next four sets of constraints (67)–(70) guarantee an appropriate break assignment to the shifts. The second approach presented in Sect. 3.2.3 is used to define break assignments. Constraints (72) and (71) force that no regular shifts can be assigned on holidays or general vacations for group j, respectively. Finally, the set (73)–(78) gives the variable definitions. For more information about the underlying problem and its formulation as a MIP see Chap. 3.

Each SP_j is defined by its unique data elements and corresponding constraints. In our computational experiments, for example, we consider full-time and part-time physicians as different groups (see Sect. 5.3). For the former, all the constraints are used to define the solution space. For the latter, just constraints (53)–(59) and (72) are active since part-timers are not given a lunch break and do not work on-call services. In the absence of the remaining constraints (60)–(71) we do not need the decision variables for breaks ($y_{j,t}^{break}$), time windows ($e_{j,w}$ and $l_{j,w}$), and on-call assignments ($y_{j,t}^{oc}$) when formulating SP_j for part-timers. Moreover, constraints (58) would be written with only the $x_{j,t}$, $o_{j,w}$, and $u_{j,w}$ variables.

If the any solution of SP_j indicates that $\overline{c}_j < 0$, then RMP does not consist of the optimal set of columns and optimal values of $\left\{ x_{j,t}, y_{j,t}^{break}, y_{j,t}^{shift}, y_{j,t}^{rest}, y_{j,l}^{oc} : \forall t \in \mathcal{T}, l \in \mathcal{L} \right\}$ are used to construct a new column k,

$$\begin{bmatrix} c_{j,k}^{sched} \\ \widehat{X}_t^{j,k} - \widehat{Y}_t^{j,k} \\ \widehat{Z}_l^{j,k} \\ \widehat{1}_j \end{bmatrix}$$

which is added to the set $\mathcal{K}(j)$ in RMP corresponding to group $j \in \mathcal{J}$. Here, $\widehat{X}_t^{j,k}$, $\widehat{Y}_t^{j,k}$, and $\widehat{Z}_l^{j,k}$ are parameter vectors that characterize the new roster that could be

54 4 Solution Methodologies

assigned to a physician in group j at a subsequent iteration. The symbol $\widehat{1}_j$ denotes a vector of length $|\mathcal{J}|$ that has a 1 at position j and a 0 at all other positions. After adding one or more columns to RMP, the latter is reoptimized and the foregoing cycle is repeated. Every SP_j solution with $\overline{c}_j < 0$ will improve the RMP solution. If $\overline{c}_j \geq 0$ for all subproblems no promising column exists and CGA terminates.

4.3.3 Finding Integer Solutions

When the linear relaxation of RMP is optimized for the first time, some form of B&B will be needed when the λ variables are not integral. Using CGA at each node of the search tree starting at the root node, provides a valid lower bound for fathoming purposes. The combination of B&B and CGA is termed B&P. If the pricing subproblems are relatively easy to solve, i.e. they have a special structure for which efficient algorithms are available, then an attempt should be made to find most negative reduced cost column (e.g., see [55, 76, 81] or [57]); if they are difficult IPs, then it might be better to terminate CGA when the first promising column is found. Various other pricing schemes are proposed in the literature like full, partial or multiple pricing (see [51]). In fact, for the approach to be valid it is only necessary to find a solution to SP_j with $\overline{c}_j < 0$ for some j and not the most negative value (e.g., see [123] or [57]).

Figure 4.1 illustrates the flow of CGA within a B&P framework. As can be seen, we conduct a starting heuristic to generate an initial set of columns (for more details to the starting heuristic see Sect. 4.3.7.1). Then we solve the RMP and based on the solution we try to find a promising column. If such a column exists we add it to RMP and re-optimize, otherwise we have found the optimal LP solution to MP. If the solution is integral at the root node we are done and report the final set of columns. If not, then we try to transform the continuous solution in a feasible solution either by a rounding heuristic (cf. Sect. 4.3.7.3) or by solving RMP as IP (see Sect. 4.3.7.2). If the new feasible solution is better then the best solution found so far, we update the *upper bound* (UB) and perform branching according to one of the strategies will be discussed shortly.

A critical aspect of B&P is the strategy selected for branching. This issue is discussed at some length by Barnhart et al. [20], Vanderbeck and Wolsey [117], Vanderbeck [116], and Wolsey [123] among others. We have developed and tested two strategies for constructing the search tree, the first based on the MP variables and herein referred to as MPVarB, and the second based on the SP variables and referred to as SPVarB. Each is described in the following sections. Beside these strategies the selection rule for the next node to explore is also a basic component of a branching strategy. There are two rules that are working in opposite direction. One is to select the node with the smallest lower bound value. Here the goal is to find a high quality feasible solution quickly. However, this approach needs much storing space, i.e. among others one has to store the objective function values, bases, and constraints that are added at previous nodes, like variable bounds for MPVarB (see Sect. 4.3.4), associated with all active nodes. A second rule, which is more

4.3 Column Generation and B&P Algorithm

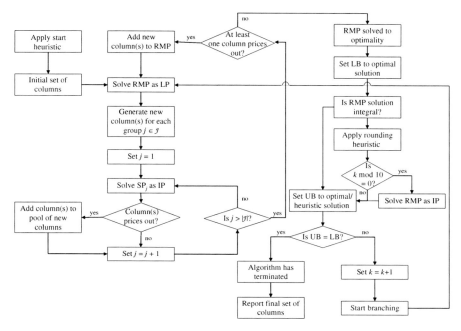

Fig. 4.1 Scheme overview CGA within B&P framework

common and much easier to implement, is to select the most recently explored node. This strategy is known as a depth-first search. Especially backtracking which is the process of moving from a fathomed node to the next node is easier to implement. The point is that the entire search tree has to be systematically explored so as to guarantee that the optimal solution is not overlooked. To limit the overall size of the search tree it is crucial to fathom nodes in an implicit enumeration scheme. Fathoming is done if the solution is integral, infeasible or the relaxed solution is worse than the best feasible solution found so far. With respect to the fathoming by bounds, the efficiency of the algorithm depends on finding high quality feasible solutions (UBs), obtaining the relaxed solutions at some nodes, also called *lower bounds* (LBs), and the general structure of the solution space. The latter means for instance that it is hard to find the optimal solution among many near optimal solutions.

4.3.4 Branching on MP Variables (MPVarB)

When CGA terminates with a fractional solution, i.e., at least one of the MP variables $\lambda_{j,k}$ is fractional, enumeration is required to achieve integrality. Let $\tilde{\lambda}_{j,k}$ for all j and k be the solution provided by CGA at some node in the search tree. A standard branching strategy for extending the tree is to create two descendant nodes by respectively enforcing $\lambda_{j,k} \leq \left\lfloor \tilde{\lambda}_{j,k} \right\rfloor$ and $\lambda_{j,k} \geq \left\lceil \tilde{\lambda}_{j,k} \right\rceil$, where $\lfloor \lambda \rfloor$ denotes the

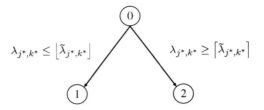

Fig. 4.2 Master problem (column) variable branching at root node

greatest integer less than or equal to λ and $\lceil \lambda \rceil$ denotes the smallest integer greater than or equal to λ. We adopt this strategy and pick the branching variable as the one whose fractional component is closest to 0.5, breaking ties arbitrarily. One can use the index k as a tiebreaker. This strategy reduces the infeasibility of the relaxed solution at most. Let (j^*, k^*) be the corresponding indices. Figure 4.2 shows the branching strategy at the root node. Note that this branching is the same type used in standard B&B for solving IPs (see [123]).

On either branch, the RMP must be modified by placing the appropriate bound on the branching variable. SP_{j^*} must also be modified to ensure that the column being excluded in the updated RMP is not regenerated when CGA is run. This is more critical for the left descendant defined by the bound $\lambda_{j^*,k^*} \leq \lfloor \tilde{\lambda}_{j^*,k^*} \rfloor$ because linear programming theory tells us that the column associated with (j^*, k^*) has a negative reduced cost and hence will be regenerated when SP_{j^*} is solved unless preventive measures are taken. To ensure that the corresponding solution is infeasible to SP_{j^*}, we add the following *roster elimination constraint* (REC) along the left branch.

$$\sum_{t \in \hat{S}^{shift}(j^*,k^*)} x_{j^*,t} + \sum_{t \in \hat{S}^{break}(j^*,k^*)} y_{j^*,t}^{break} + \sum_{t \in \hat{\mathcal{L}}(j^*,k^*)} y_{j^*,t}^{oc} + \alpha_{j^*,k^*} - \left(\sum_{t \in S^{shift}(j^*,k^*)} x_{j^*,t} + \sum_{t \in S^{break}(j^*,k^*)} y_{j^*,t}^{break} + \sum_{t \in \mathcal{L}(j^*,k^*)} y_{j^*,t}^{oc} \right) \geq 1 \quad (79)$$

where $\alpha_{j^*,k^*} = |S^{shift}(j^*, k^*)| + |S^{break}(j^*, k^*)| + |\mathcal{L}(j^*, k^*)|$. The sets $S^{shift}(j^*, k^*)$, $S^{break}(j^*, k^*)$, and $\mathcal{L}(j^*, k^*)$ corresponding to j^* and k^* and their complements $\hat{S}^{shift}(j^*, k^*)$, $\hat{S}^{break}(j^*, k^*)$, and $\hat{\mathcal{L}}(j^*, k^*)$ are defined as follows

$$S^{shift}(j^*, k^*) = \{t \in T^{dem} : X_t^{j^*,k^*} = 1\}$$
$$S^{break}(j^*, k^*) = \{t \in T^{dem} : Y_t^{j^*,k^*} = 1\}$$
$$\mathcal{L}(j^*, k^*) = \{t \in \mathcal{L} : Z_t^{j^*,k^*} = 1\}$$

4.3 Column Generation and B&P Algorithm

$$\hat{\mathcal{S}}^{shift}\left(j^*,k^*\right) = \mathcal{T}^{dem} \setminus \mathcal{S}^{shift}\left(j^*,k^*\right)$$
$$\hat{\mathcal{S}}^{break}\left(j^*,k^*\right) = \mathcal{T}^{dem} \setminus \mathcal{S}^{break}\left(j^*,k^*\right)$$
$$\hat{\mathcal{L}}\left(j^*,k^*\right) = \mathcal{L} \setminus \mathcal{L}\left(j^*,k^*\right).$$

In the case of the right branch, which is defined by the bound $\lambda_{j^*,k^*} \geq \left\lceil \tilde{\lambda}_{j^*,k^*} \right\rceil$, no modifications to SP_{j^*} are necessary. Informally, we can see this by observing that the solution to RMP calls for the use of λ_{j^*,k^*} physicians in group j^* with roster k^*. Because the partition calls for the use of at least $\left\lceil \tilde{\lambda}_{j^*,k^*} \right\rceil$ physicians in this group, which is more than the optimum $\tilde{\lambda}_{j^*,k^*}$, SP_{j^*} will never produce a solution that corresponds to column (j^*,k^*). The reduced cost of such a column will also be ≥ 0.

For example, assuming a ten period planning horizon with the following assignment vector $\hat{X}^k$ where k is the branching variable. Table 4.4 shows the example settings. To facilitate the presentation we do not consider different physician groups, on-call services, and break assignments but the example can easily be extended to more complex case. When creating a left branch we forbid SP from re-generating column $\hat{X}^k$ by imposing REC on SP. First, we calculate α_k which is the sum of all the entries contained in $\hat{X}^k$. Here α_k is 5. Second, we collect all the x-variables that correspond to a 1 in the vector $\hat{X}^k$. We denote this summation by $term_1$ (see Table 4.4). Then, we collect all the x-variables that correspond to a 0 in the assignment vector which is denoted by $term_2$. Finally we state REC as $term_2 + 5 - term_1 \geq 1$. If SP generates the same roster then $term_1 = 5$ and $term_2 = 0$ but in this case REC is violated since the left-hand side is 0. Hence, either a variable in $term_1$ has to be 0 or a variable in $term_2$ has to be 1 and this would result in a different roster.

The advantage of MPVarB is its simplicity; its disadvantage is that it leads to an unbalanced partitioning of the feasible region. On the left branch, only one pattern is excluded so the updated RMP is only marginally more restricted than the RMP at its predecessor node. On the right branch, however, we are required to assign at least $\left\lceil \tilde{\lambda}_{j^*,k^*} \right\rceil$ physicians in group j^* to roster k^*. Given that the parent RMP may have no restrictions on λ_{j^*,k^*}, the feasible region associated with the right branch is likely to be much tighter than that of its parent; that is, $\left\lceil \tilde{\lambda}_{j^*,k^*} \right\rceil$ fewer physicians are available for assignments to rosters other than k^*.

Table 4.4 Example REC

Period	1	2	3	4	5	6	7	8	9	10
$\hat{X}_t^k$	0	1	1	1	0	1	1	0	0	0
$term_1$		x_2	$+x_3$	$+x_4$		$+x_6$	$+x_7$			
$term_2$	x_1				$+x_5$			$+x_8$	$+x_9$	$+x_{10}$

4.3.5 Branching on SP Variables (SPVarB)

A second approach for branching is based on the subproblem variables and is relatively straightforward to implement when the MP variables are binary, the overwhelming situation in which column generation has been used. One popular strategy has been introduced by Ryan and Foster [102] for pure set covering and set partitioning problems. When the MP variables are integer, as it is the case here, Vanderbeck and Wolsey [117] extended the Ryan and Foster scheme to allow each element of the right-hand side vector in MP to be a nonnegative integer. Our strategy, denoted by SPVarB, makes use of this ideas.

Given the relationship between the SP_j variables and the RMP variables, if RMP generates a fractional solution $\lambda_{j,k}$ then there exist at least one subset $\mathcal{F}(j) \subseteq \mathcal{T}^{dem} \cup \mathcal{L}$ with a fractional sum of λ's for at least one physician group j. The subset $\mathcal{F}(j)$ is called a branching pattern and its length is defined as its cardinality $|\mathcal{F}(j)|$. A column $k \in \mathcal{K}(j)$ satisfies branching pattern $\mathcal{F}(j)$ if $X_t^{j,k} - Y_t^{j,k} = 1$ for $t \in \mathcal{F}(j) \cap \mathcal{T}^{dem}$ and $Z_t^{j,k} = 1$ for $t \in \mathcal{F}(j) \cap \mathcal{L}$. As such, if no such subset exist then all λ's have to be integral. To implement this result, define a $\beta_{\mathcal{F}(j)}$ for each subset $\mathcal{F}(j)$ as follows:

$$\beta_{\mathcal{F}(j)} \equiv \sum_{k \in \mathcal{K}(j, \mathcal{F}(j))} \lambda_{j,k}, \forall j \in \mathcal{J}, \mathcal{F}(j) \in \mathcal{B} \tag{80}$$

where $\mathcal{B}$ is the set of all possible subsets of $\mathcal{F}(j)$ for all $j \in \mathcal{J}$.

Although the cardinality of $\mathcal{B}$ grows exponentially with the number of periods in the planning horizon, $|\mathcal{T}|$, and hence is extremely large, it is sufficient to find one subset $\mathcal{F}(j)$ for one group j where β is fractional. The most efficient way to do this is to start by investigating all subsets of $\mathcal{F}(j)$ of length 1; that is, $|\mathcal{F}(j)| = 1$, $\forall j \in \mathcal{J}$. If no fractional β can be found then all patterns with $|\mathcal{F}(j)| = 2$ are examined, and so on until a fractional β is found or no such subset $\mathcal{F}(j)$ exists for any physician group $j \in \mathcal{J}$.

We now give a small example that illustrates the branching. To facilitate the presentation we consider just one physician type in the problem so we can omit the index j. Table 4.5 gives an LP solution at some node where eight rosters are contained in RMP and some of the λ-variables are fractional. Tables 4.6–4.8 show the values of β for patterns of different cardinality. A "1-pattern", for example, corresponds to a single row. For row 1, adding the fractional λ-values gives 5 (columns 1, 2, 4, 6).

The first and only fractional β is found when $|\mathcal{F}| = 3$ corresponding to a 3-pattern branching strategy where $\mathcal{F}$ consists of periods 1, 2 and 4. In the example the cardinality of $\mathcal{B}$ is 15. When we examine the 1- and 2-patterns we do not find a fractional β for branching. Recall, the Ryan and Foster scheme just considers 2 pattern branching but this would not eliminate the fractional solution in the example. Hence, our branching can handle the more general case for set covering formulations with right-hand sides that are general integer.

4.3 Column Generation and B&P Algorithm

Table 4.5 Example branching strategy SPVarB

t	k (Rosters)								d_t
	1	2	3	4	5	6	7	8	(RHS)
1	1	1	0	1	0	1	0	0	5
2	0	1	1	1	0	0	0	1	4
3	0	0	0	0	1	0	0	0	2
4	0	1	1	0	0	1	1	0	6
	2.5	0.5	2.5	0.5	2.0	1.5	1.5	0.5	
				λ_k					

Table 4.6 1-pattern

Row	1	2	3	4
β	5	4	2	6

Table 4.7 2-pattern

Rows	1-2	1-3	1-4	2-3	2-4	3-4
β	1	0	2	0	3	0

Table 4.8 3-pattern

Rows	1-2-3	1-2-4	1-3-4	2-3-4
β	0	0.5	0	0

Since the β-values have to be general integer at optimality an obvious branching rule is to force them to be integral as a way of creating the search tree. As with MPVarB, we choose the branching variable to be the β that is most fractional and create two descendant nodes. The first includes the constraint (81) below and the second includes (82), where $\mathcal{K}(j, \mathcal{F}(j))$ is the set of columns in RMP associated with group j that exhibit pattern $\mathcal{F}(j)$.

$$\sum_{k \in \mathcal{K}(j,\mathcal{F}(j))} \lambda_{j,k} \leq \lfloor \beta_{\mathcal{F}(j)} \rfloor \tag{81}$$

$$\sum_{k \in \mathcal{K}(j,\mathcal{F}(j))} \lambda_{j,k} \geq \lceil \beta_{\mathcal{F}(j)} \rceil \tag{82}$$

Figure 4.3 shows the branching strategy at the root node for an arbitrary $\mathcal{F}(j)$ and j corresponding to the pattern and the physician type we choose for branching.

On the left branch, at most $\lfloor \beta_{\mathcal{F}(j)} \rfloor$ physicians from group j will be permitted to work pattern $\mathcal{F}(j)$, while on the right branch, at least $\lceil \beta_{\mathcal{F}(j)} \rceil$ physicians will be required to work pattern $\mathcal{F}(j)$. To impose these restrictions in SP_j we set

$$\sum_{t \in \mathcal{F}(j) \cap T^{dem}} \left(x_{j,t} - y_{j,t}^{break} \right) + \sum_{t \in \mathcal{F}(j) \cap \mathcal{L}} y_{j,t}^{oc} \leq |\mathcal{F}(j)| - 1 \tag{83}$$

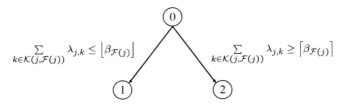

Fig. 4.3 Subproblem (original) variable branching at root node

for the left branch and

$$\sum_{t \in \mathcal{F}(j) \cap T^{dem}} \left(x_{j,t} - y_{j,t}^{break} \right) + \sum_{t \in \mathcal{F}(j) \cap \mathcal{L}} y_{j,t}^{oc} = |\mathcal{F}(j)| \cdot y_{j,n}^{ind} \quad (84)$$

for the right branch. Constraint (83) prevents more than $|\mathcal{F}(j)| - 1$ of the binary variables that define $\mathcal{F}(j)$ from being in a solution to SP$_j$, while (84) requires that the solution covers either all periods $t \in \mathcal{F}(j)$ or none of them. To enforce the latter requirement, we introduce a new binary variable, $y_{j,n}^{ind}$, that takes the value of 1 when all periods in $\mathcal{F}(j)$ are covered by a solution to SP$_j$, and 0 if none are covered.

Adding new constraints to RMP during branching introduces an equivalent number of dual variables that have to be included in the objective function (52) of SP$_j$. Let $\delta_{j,n}$ be the new dual variable for physician group j at node n created by imposing one of the constraints (81) or (82) on RMP. For the left branch, $\delta_{j,n} \leq 0$, and for the right, $\delta_{j,n} \geq 0$. The updated objective function of SP$_j$ is

$$\overline{c}_j = \sum_{w \in \mathcal{W}} \left(c_j^{paid} \cdot h_{j,w} + c_j^{over} \cdot o_{j,w} \right) - \sum_{t \in T^{dem}} \delta_t^{dem} \cdot \left(x_{j,t} - y_{j,t}^{break} \right)$$
$$- \sum_{t \in \mathcal{L}} \delta_t^{oc} \cdot y_{j,t}^{oc} - \delta_j^{conv} - \sum_{i \in \mathcal{N}(n)} \omega_{j,i} \cdot \delta_{j,i} \cdot y_{j,i}^{ind} \quad (85)$$

where the set $\mathcal{N}(n)$ contains all the nodes connecting the root node with node n and $\omega_{j,i}$ is an indicator parameter equal to 1 if a right descendant is created at node $i \in \mathcal{N}(n)$, 0 otherwise. Now, because we prohibit regenerating pattern $\mathcal{F}(j)$ when creating a left descendant, the respective dual variables are not included in (85) (i.e. $\omega_{j,i} = 0$). When creating a right descendant, though, they are included and the logic variable $y_{j,i}^{ind}$ decides whether or not the new column covers all elements in $\mathcal{F}(j)$. In addition, when SP$_j$ produces a promising roster, say, at node n, the right master problem branching constraints (82), $\sum_{k \in \mathcal{K}(j, \mathcal{F}(j))} \lambda_{j,k} \geq \lceil \beta_{\mathcal{F}(j)} \rceil$, at all nodes on the path from the root node to node n, i.e., for all $i \in \mathcal{N}(n)$, have to be updated when the corresponding indicator variables $y_{j,i}^{ind} = 1$, $i \in \mathcal{N}(n)$ meaning that the new λ variable is added to the left hand side. This observation is more visible by looking at the dual of RMP.

4.3.6 A Dual Point of View

Every MP has a dual, the *dual master problem* (DMP) that is implicitly solved during column generation. In particular, the dual of (45)–(49) is given as follows. We use the same notation as previously stated (see Sect. 4.3.1).

Dual master problem

$$\text{Max} \sum_{t \in \mathcal{T}^{dem}} d_t \cdot \delta_t^{dem} + \sum_{t \in \mathcal{L}} d_t^{oc} \cdot \delta_t^{oc} + \sum_{j \in \mathcal{J}} n_j^{phy} \cdot \delta_j^{conv} \tag{86}$$

subject to

$$\sum_{t \in \mathcal{T}^{dem}} \left(X_t^{j,k} - Y_t^{j,k} \right) \cdot \delta_t^{dem} + \sum_{t \in \mathcal{L}} Z_t^{j,k} \cdot \delta_t^{oc} + \delta_j^{conv} \le c_{j,k}^{sched},$$
$$\forall j \in \mathcal{J}, k \in \mathcal{K}(j) \tag{87}$$

$$\delta_t^{dem} \le c^{out}, \forall t \in \mathcal{T}^{dem} \tag{88}$$

$$\delta_t^{oc} \le c^{oc\_out}, \forall t \in \mathcal{L} \tag{89}$$

$$\delta_t^{dem}, \delta_l^{oc} \ge 0, \delta_j^{conv} \le 0, \forall t \in \mathcal{T}^{dem}, l \in \mathcal{L}, j \in \mathcal{J} \tag{90}$$

The process of searching for a new column k for RMP by solving SP_j produces a new constraint of the form (87) in DMP when $\overline{c}_{j,k} < 0$ at optimality. This constraint represents a cut in the dual solution space. From LP theory, we know that constraints (87) are tight for basic variables $\lambda_{j,k}$ but may be loose otherwise. If we can find a new column with negative reduced cost the corresponding cut (87) is added to DMP and RMP is reoptimized. Otherwise, we have the optimal solution to the linear RMP. Subsequently we show the changes in the dual solution space when either MPVarB (cf. Sect. 4.3.4) or SPVarB (cf. Sect. 4.3.5) is employed.

4.3.6.1 Dual Space by MPVarB

When MPVarB is performed, simple bounds are placed on those variables in RMP that were selected for branching. The dual variables associated with these bounds appear in (87) on the left-hand side for an already existing column in RMP and so alter the dual feasible region, making it either looser when an upper bound on $\lambda_{j,k}$ is imposed or tighter when a lower bound is imposed. The new dual variables are denoted by $\delta_{j,k}^{UB}$ or $\delta_{j,k}^{LB}$ depending whether an upper bound or a lower bound on some $\lambda_{j,k}$ is introduced. Note that $\delta_{j,k}^{UB} \le 0$ and $\delta_{j,k}^{LB} \ge 0$ holds. In case an upper bound is imposed in RMP, $\lambda_{j^*,k^*} \leq \lfloor \tilde{\lambda}_{j^*,k^*} \rfloor$, where j^* and k^* corresponds to the branching variable, constraint (87) changes to

$$\sum_{t \in \mathcal{T}^{dem}} \left(X_t^{j^*,k^*} - Y_t^{j^*,k^*} \right) \cdot \delta_t^{dem}$$

$$+ \sum_{t \in \mathcal{L}} Z_t^{j^*,k^*} \cdot \delta_t^{oc} + \delta_{j^*}^{conv} + \delta_{j^*,k^*}^{UB} \leq c_{j^*,k^*}^{sched} \quad (91)$$

Furthermore, the objective function of DMP changes to

$$\text{Max} \sum_{t \in \mathcal{T}^{dem}} d_t \cdot \delta_t^{dem} + \sum_{t \in \mathcal{L}} d_t^{oc} \cdot \delta_t^{oc} + \sum_{j \in \mathcal{J}} n_j^{phy} \cdot \delta_j^{phy} + \lfloor \tilde{\lambda}_{j^*,k^*} \rfloor \cdot \delta_{j^*,k^*}^{UB} \quad (92)$$

As can be seen these new dual variables associated with bounds of existing columns are not added to the objective function of SP_j when a new promising column is searched since they are just valid for a specific j and k associated with the branching (cf. Sect. 4.3.4). But recall that for each left branch an REC is added to the feasible region (see Sect. 4.3.4). As a consequence, when SP_j is re-solved at the current node, columns that were absent at the parent node might now be generated for RMP. New columns in RMP translate into new cuts (87) for DMP.

4.3.6.2 Dual Space by SPVarB

When SPVarB is used, a new column produces the following cut in DMP.

$$\sum_{t \in \mathcal{T}^{dem}} \left(X_t^{j^*,k^*} - Y_t^{j^*,k^*} \right) \cdot \delta_t^{dem} + \sum_{t \in \mathcal{L}} Z_t^{j^*,k^*} \cdot \delta_t^{oc} + \delta_j^{conv}$$

$$+ \sum_{i \in \mathcal{N}(n)} \omega_{j,i} \cdot Y_{j,i}^{ind} \cdot \delta_{j,i} \leq c_{j,k}^{sched} \quad (93)$$

As mentioned, when creating a left branch we impose constraint (83), $\sum_{t \in \mathcal{F}(j) \cap \mathcal{T}^{dem}} \left(x_{j,t} - y_{j,t}^{break} \right) + \sum_{t \in \mathcal{F}(j) \cap \mathcal{L}} y_{j,t}^{oc} \leq |\mathcal{F}(j)| - 1$ on SP_j at some node n and (87) does not change since the new column does not cover the periods in $\mathcal{F}(j)$. Therefore, the new column is not added to the left hand side of constraint (81) when node i is a left descendant (i.e. $\omega_{j,i} = 0$). When a right descendant is generated (i.e. $\omega_{j,i} = 1$) the new column covers all the periods in $\mathcal{F}(j)$ or none of them. This implies that the nonnegative dual value $\delta_{j,i}$ is present in (93) when the indicator parameter $Y_{j,i}^{ind} = 1$, which is equivalent to $y_{j,i}^{ind} = 1$ in SP_j. Constraint (93) is essentially the objective function given in (83) when SPVarB is employed. Adding new constraints of the form (81) or (82) to RMP change the objective function (86) of DMP accordingly to

4.3 Column Generation and B&P Algorithm 63

$$\text{Max} \sum_{t \in T^{dem}} d_t \cdot \delta_t^{dem} + \sum_{t \in \mathcal{L}} d_t^{oc} \cdot \delta_t^{oc} + \sum_{j \in \mathcal{J}} n_j^{phy}$$
$$+ \sum_{i \in \mathcal{N}(n)} \left((1 - \omega_{i,j}) \cdot \lfloor \beta_{\mathcal{F}(j)} \rfloor + \omega_{i,j} \cdot \lceil \beta_{\mathcal{F}(j)} \rceil \right) \cdot \delta_{j,i} \quad (94)$$

4.3.7 Heuristics for the B&P Algorithm

In this part of the thesis we present some heuristic procedures that improve the performance of the B&P algorithm. To maximize the efficiency of any B&P algorithm it is crucial to have a procedure that provides a good set of initial columns. In addition, some method for obtaining integer solutions during the enumeration process should be included to increase the opportunity for fathoming. We will present two methods to find integer solutions in Sect. 4.3.7.2 and in Sect. 4.3.7.3, respectively.

4.3.7.1 Initialization Heuristic

Our initialization heuristic is divided in two parts. First, we construct a nominal set of rosters whose shifts each include proper break assignments. Starting with the first period in which there is positive demand in each week in the planning horizon, we define three 9-h shifts. The breaks are assigned a fixed number of periods after the starting period each day and are laddered: first, 3 periods after the starting period, then 4 periods, and then 5 periods. For example, for a 1-week planning horizon, if the first demand period in the week is period 8, then we construct a Monday to Friday roster starting in period 8 and having a break in period 11. The shifts defining the second and third rosters also start in period 8 but have breaks in periods 12 and 13, respectively. The next three rosters all start one period later (period 9 in the example) and have breaks in periods 12, 13, and 14, respectively. The last three rosters start two periods later (period 10, here), and have identical break patterns, with the first starting three periods into a shift. If there are 2 weeks in the planning horizon and the earliest shift in the second week is period 13, we would first extend the original 9 rosters through the second week with duplicate patterns (the first 3 would start in period 8 and so on). We would then construct 9 more rosters with the first starting in period 13 of the first week and having a break in period 16. If there are $|\mathcal{W}|$ weeks in the planning horizon, then this process would lead to $9 \cdot |\mathcal{W}|$ initial rosters that reflected the above logic.

For the hospital under study, on-call service begins in period 8 each day of the planning horizon. In addition to the rosters already constructed, we construct $7 \cdot |\mathcal{W}| + 7 \cdot (|\mathcal{W}| - 1)$ more rosters with on-call service assignments using shifts that all start in period 8 and have a break in period 11 as in the previous example. The first $7 \cdot |\mathcal{W}|$ rosters have a single on-call service on successive days of the week. The next $7 \cdot |\mathcal{W}|$ have two on-call services. The first of these has an on-call service on Monday of the first week and Monday of the second week. The second has an on-call service on Tuesday of the first week and Tuesday of the second week, and so on. The next $7 \cdot |\mathcal{W}|$ rosters have three on-call services. The first has an on-call service on Monday of the first 3 weeks, the second on Tuesday of the first 3 weeks and so on. Figure 4.4 shows exemplarily the construction of part 1 rosters for a 2-week planning horizon with earliest demand period of 8 in the first week and 13 in the second week. In total, 39 rosters are generated where the first nine are based on the first week, the next nine are based on the second week demand input, and the remaining rosters have on-call assignments. Periods 49–336 are omitted in Fig. 4.4 to facilitate the presentation. Again the assignment on Monday in the first week is identical to all assignments on any working day (Monday through Friday) in the planning horizon except when an on-call assignment takes place in the schedule. A 0 signals that the physician is off duty because of a break assignment or a rest period.

In the second part, rosters are constructed based on the specific problem instance. Since the feasible region of SP_j embodies the rules that all group j rosters must satisfy, we use it to identify feasible columns. The procedure starts by replacing (52) with the following objective function.

$$
\text{Min} \sum_{w \in \mathcal{W}} \left(c_j^{paid} \cdot h_{j,w} + c_j^{over} \cdot o_{j,w} \right)
$$
$$
- \sum_{t \in \mathcal{T}^{dem}} d_t \cdot \left(x_{j,t} - y_{j,t}^{break} \right) - \sum_{t \in \mathcal{L}} d_t^{oc} \cdot y_{j,t}^{oc} \quad (95)
$$

Then, for $j = 1$, the resultant problem is solved using the original demand d_t for $t \in \mathcal{T}^{dem}$ and d_t^{oc} for $t \in \mathcal{L}$ to get a new roster which is converted to a column and added to RMP. The computations are terminated when either the optimum is found or a 10 s time limit is reached. At that point, the demand is reduced by 1 in each period covered by the new roster, i.e. $d_t \leftarrow d_t - \left(x_{j,t} - y_{j,t}^{break} \right) \ \forall t \in \mathcal{T}^{dem}$ and $d_t^{oc} \leftarrow d_t^{oc} - y_{j,t}^{oc} \ \forall t \in \mathcal{L}$, and the problem is re-solved to get a second roster. The process is repeated until all d_t and d_t^{oc} are 0 or a predefined number of columns, e.g. equal to the number of physicians in group 1, n_1^{phy}, are generated. When one of these conditions is met, we put $j \leftarrow j + 1$, reset d_t and d_t^{oc} to their original values, and repeat the process described for all $j \in \mathcal{J}$.

4.3.7.2 Feasibility Heuristic to Find Integer Solutions

To obtain feasible solutions to the original problem and hence upper bounds we solve the RMP as an IP at every 10 nodes of the search tree. Before doing so, though, we remove all branching constraints on the path from the root node to the current node. This increases the runtime, especially when RMP consists of a large number of columns. However, this strategy worked best in our preliminary testing. At the root node, we get our first incumbent; subsequently, whenever the IP solution provides a better objective function value, we update the incumbent (new UB) and fathom all nodes whose LB is greater than or equal to the new incumbent. Other options

4.3 Column Generation and B&P Algorithm

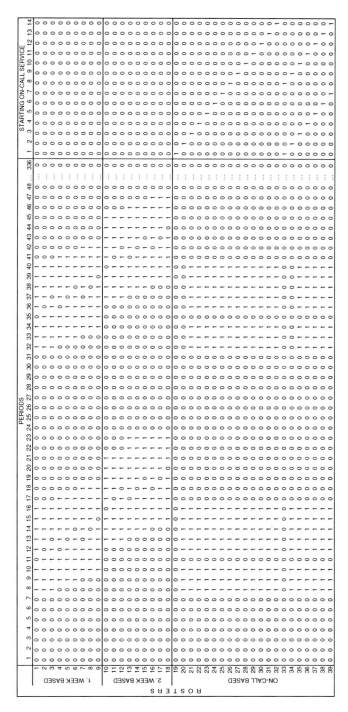

Fig. 4.4 Example for initialization heuristic with 2-week planning horizon

for finding feasible solutions include variable fixing and then solving the remaining RMP as an IP (e.g., see [99]), or the use of metaheuristics and rounding heuristics to convert continuous solutions into integer solutions at various nodes in the search tree. One such rounding heuristic is presented in the next section.

4.3.7.3 Rounding Heuristic to Find Integer Solutions

As mentioned before, another option to find feasible solutions is to apply some sort of rounding. In preliminary developments of the rounding heuristic we set all fractional solutions to their closest integer value. Even if it does not cost much computation time the strategy has been only in few cases successful to find feasible solutions (new UBs) that helped terminating the algorithm. We then enhance the strategy as follows. First, we round up all fractional solutions for each group $j \in \mathcal{J}$ to the next integer value then we determine the infeasibility of the convexity constraint (48) in MP for each group j as $\Delta_j = \sum_{i \in \mathcal{K}(j)} \lceil \lambda_{j,k} \rceil - |\mathcal{I}_j|$ $\forall j \in \mathcal{J}$. Furthermore, we sort the λ's according to the fractional part of the solution in ascending order. Finally, we reduce the first Δ_j entries by 1 to achieve feasibility in each group $j \in \mathcal{J}$ and determine the coverage. To calculate the cost of the final assignments we weight all uncovered periods or on-call services by their associated cost values, c^{out} or c^{oc_out}, respectively, and add the cost for the assigned schedules, $c^{sched}_{j,k}$. The pseudo code of the rounding heuristic is given in Fig. 4.5.

Subsequently we give a small example that illustrates the rounding heuristic. To facilitate the presentation we consider just one group of physicians j. However, the example can be extended to the more general case where different physician groups are considered. Given a assignment with three fractional solutions,

Input:	Fractional solution $\lambda_{j,k}$ $\forall j \in \mathcal{J}, i \in \mathcal{K}(j)$ and the number of physicians $|\mathcal{I}_j|$ for each group j.
Output:	Feasible solution containing a final set of assignments with associated cost.
Step 0:	Round up all fractional $\lambda_{j,k}$ to the next integer value $\lambda^{int}_{j,k}$ with $\lambda^{int}_{j,k} = \lceil \lambda_{j,k} \rceil$ $\forall j \in \mathcal{J}, i \in \mathcal{K}(j)$
Step 1:	Determine the infeasibility of the solution Δ_j for each group j with $\Delta_j = \sum_{i \in \mathcal{K}(j)} \lambda^{int}_{j,k} - |\mathcal{I}_j|$ $\forall j \in \mathcal{J}$
Step 2:	Sort all $\lambda^{int}_{j,k}$ in ascending order according to the fractional part $\lambda^{frac}_{j,k}$ of the solution with $\lambda^{frac}_{j,k} = \lambda_{j,k} - \lfloor \lambda_{j,k} \rfloor$ $\forall j \in \mathcal{J}, i \in \mathcal{K}(j)$
Step 3:	If $\Delta_j > 0$ then replace the first Δ_j entries according to the sorted set with condition $\Delta_j > 0$ by $\lambda^{int}_{j,k} = \lambda^{int}_{j,k} - 1$
Step 4:	Determine number of under coverage in demand periods $t \in \mathcal{T}^{dem}$ with $\Delta^{dem}_t = \max \left(d_t - \sum_{j \in \mathcal{J}} \sum_{k \in \mathcal{K}(j)} \left(X^{j,k}_t - Y^{j,k}_t \right) \cdot \lambda^{int}_{j,k}, 0 \right)$, $\forall t \in \mathcal{T}^{dem}$
Step 5:	Determine number of under coverage for an on-call services starting in period $t \in \mathcal{L}$ with $\Delta^{oc}_t = \max \left(d^{oc}_t - \sum_{j \in \mathcal{J}} \sum_{k \in \mathcal{K}(j)} Z^{j,k}_t \cdot \lambda^{int}_{j,k}, 0 \right)$, $\forall t \in \mathcal{L}$
Step 6:	Calculate the final cost of the assignments c^{fin} as $c^{fin} = \sum_{j \in \mathcal{J}} \sum_{k \in \mathcal{K}(j)} c^{sched}_{j,k} \cdot \lambda^{int}_{j,k} + \sum_{t \in \mathcal{T}^{dem}} c^{out} \cdot \Delta^{dem}_t + \sum_{t \in \mathcal{L}} \Delta^{oc}_t \cdot x^{oc_{out}}_t$
Step 7:	Report feasible assignments $\lambda^{int}_{j,k}$ and final cost c^{fin}.

Fig. 4.5 Pseudo code rounding heuristic

4.3 Column Generation and B&P Algorithm

Table 4.9 Example of the rounding heuristic

	Input		Step 0, 1		Step 2	Step 3, 6, 7
k	c_k	λ_k	λ_k^{int}	λ_k^{frac}	sort. list pos.	λ_k^{int}
1	11	0.6	$\rightarrow$ 1	0.6	$\rightarrow$ 4	1
2	0	1.1	$\rightarrow$ 2	0.1	$\rightarrow$ 2	1
3	0	2.3	$\rightarrow$ 3	0.3	$\rightarrow$ 3	2
4	0	1	$\rightarrow$ 1	0	$\rightarrow$ 1	1
		$|\mathcal{I}| = 5$		$\Delta = 2$		$c^{fin} = 11$

$\lambda_1 = 0.6$, $\lambda_2 = 1.1$, $\lambda_3 = 2.3$, and one integral solution, $\lambda_4 = 1$ and the number of available physicians (i.e. $|\mathcal{I}| = 5$). We start by rounding up all fractional λ (*Step 0*), then determining the infeasibility (*Step 1*), resorting the λ's, and transforming the infeasibility into feasibility (*Step 2*). The process is shown in Table 4.9. Once we have a feasible solution we identify the under coverage for the demand periods and on-call assignments to calculate the final cost. We omit *Step 4* and *Step 5* here since we assume that the under coverage for both is 0. So, just the costs of the assigned schedules are relevant. With $c_1^{sched} = 11$ and $c_2^{sched} = c_3^{sched} = c_4^{sched} = 0$ and the final assignment given in the last column of Table 4.9 we obtain total cost of 11.

4.3.8 Enhancements for the B&P Algorithm

We now present some ideas to improve the performance of CGA embedded in the B&P framework. First, we show the computation of lower bounds, thereon we give a procedure for early termination of the algorithm in Sect. 4.3.8.1. Second, we present a method to aggregate different physician groups j into a super physician group (see Sect. 4.3.8.2). That might be beneficial with respect to the computational burden since we do not have to solve a series of SP_j's but just the new aggregated SP.

4.3.8.1 Lower Bounds and Early Termination

In a B&P approach LBs are important which will be discussed subsequently. The first one we will present is a valid lower bound LB^{IP} on the optimal solution value of the original IP which when it is tight, helps in keeping the search tree small. The linear programming solution (LB^{LP}) would provide such a lower bound. Consequently, if the cost structure is general integer, which can always be achieved by an appropriate scaling, then $LB^{IP} = \lceil LB^{LP} \rceil$. Furthermore, adding valid inequalities to strengthen the lower bound provided by the linear programming relaxation is a commonly know standard technique. Assuming a resource assignment problem with the primary objective of minimizing the number of resources needed to execute all tasks Dumas et al. [61] presented valid inequalities to improve on the lower bound. When such valid inequalities are added to a relaxed solution, new columns might price out favorable to enter the basis. One distinctive advantage of this procedure is that the computational effort to solve the subproblems does not increase since the new dual variables appear as constants in the objective function (52) which is a desirable feature. In our case, the modified valid inequalities are as follows:

$$\sum_{j \in \mathcal{J}} \sum_{k \in \mathcal{K}(j)} \lambda_{j,k} \geq \left\lceil \sum_{j \in \mathcal{J}} \sum_{k \in \mathcal{K}(j)} \tilde{\lambda}_{j,k} \right\rceil. \tag{96}$$

Note that optimizing over MP would provide a valid LB^{LP} – the optimal solution to MP denoted with z^{MP}. However, we are optimizing over RMP which contains just a small subset of all possible columns. So, we get a optimal solution to RMP at each iteration u of CGA depending on the columns already included in RMP. We indicate this solution with z_u^{RMP}. Note that $z_u^{RMP} \geq z^{MP}$ holds for each u. Hence, z_u^{RMP} does not provide a valid lower bound in order to terminate CGA. In the following we show how to achieve lower bounds in a column generation framework. Let LB_u be a valid lower bound on z_u^{RMP} at each iteration u and let LB_n denote a valid lower bound at node n in the search tree. When we explore a new node we initialize LB_u with LB_n from the immediate predecessor node. After each iteration u in CGA we update $\text{LB}_u \leftarrow \max\{\text{LB}_u, \text{LB}_{u-1}\}$. Afterwards we present a formula to calculate LB_u at each iteration of CGA (cf. [84]). Given the current RMP solution, z_u^{RMP}, at iteration u, the corresponding dual solution, $\left\{\delta_t^{dem}, \delta_l^{oc}, \delta_j^{conv} : t \in \mathcal{T}^{dem}, l \in \mathcal{L}, j \in \mathcal{J}\right\}$, and the optimal solutions to the j pricing subproblems denoted by k_j^* in terms of MP notation with

$$\begin{aligned}
\overline{c}_{j,k_j^*} &= c_{j,k_j^*}^{sched} - \sum_{t \in \mathcal{T}^{dem}} \delta_t^{dem} \cdot \left(X_t^{j,k_j^*} - Y_t^{j,k_j^*}\right) - \sum_{t \in \mathcal{L}} \delta_t^{oc} \cdot Z_t^{j,k_j^*} - \delta_j^{conv} \\
&\leq c_{j,k}^{sched} - \sum_{t \in \mathcal{T}^{dem}} \delta_t^{dem} \cdot \left(X_t^{j,k} - Y_t^{j,k}\right) \\
&\quad - \sum_{t \in \mathcal{L}} \delta_t^{oc} \cdot Z_t^{j,k} - \delta_j^{conv}, \forall k \in \mathcal{K}(j)
\end{aligned}$$

where $\mathcal{K}(j)$ indicates all possible schedules produced by SP_j then we can compute LB_u as follows

$$\text{LB}_u = z_u^{RMP} + \sum_{j \in \mathcal{J}} \overline{c}_{j,k_j^*} \cdot n_j^{phy}. \tag{97}$$

Suppose that λ is a feasible solution to MP then (97) holds. This can be shown as follows:

$$\sum_{j \in \mathcal{J}} \sum_{k \in \mathcal{K}(j)} c_{j,k}^{sched} \cdot \lambda_{j,k} + \sum_{t \in \mathcal{T}^{dem}} c^{out} \cdot x_t^{out} + \sum_{t \in \mathcal{L}} c^{oc\_out} \cdot x_t^{oc\_out}$$

$$\geq \sum_{j \in \mathcal{J}} \sum_{k \in \mathcal{K}(j)} \left(\overline{c}_{j,k_j} + \sum_{t \in \mathcal{T}^{dem}} \delta_t^{dem} \cdot \left(X_t^{j,k} - Y_t^{j,k}\right)\right.$$

$$+ \sum_{t \in \mathcal{L}} \delta_t^{oc} \cdot Z_t^{j,k} + \delta_j^{conv} \Bigg) \cdot \lambda_{j,k}$$

$$+ \sum_{t \in \mathcal{T}^{dem}} \delta_t^{dem} \cdot x_t^{out} + \sum_{t \in \mathcal{L}} \delta_t^{oc} \cdot x_t^{oc\_out}$$

$$= \sum_{j \in \mathcal{J}} \left(\overline{c}_{j,k_j} + \delta_j^{conv} \right) \cdot \sum_{k \in \mathcal{K}(j)} \lambda_{j,k}$$

$$+ \sum_{t \in \mathcal{T}^{dem}} \delta_t^{dem} \cdot \left(\sum_{j \in \mathcal{J}} \sum_{k \in \mathcal{K}(j)} \left(X_t^{j,k} - Y_t^{j,k} \right) \cdot \lambda_{j,k} + x_t^{out} \right)$$

$$+ \sum_{t \in \mathcal{L}} \delta_t^{oc} \cdot \left(\sum_{j \in \mathcal{J}} \sum_{k \in \mathcal{K}(j)} Z_t^{j,k} \cdot \lambda_{j,k} + x_t^{out} \right)$$

$$\geq \sum_{j \in \mathcal{J}} \left(\overline{c}_{j,k_j^*} + \delta_j^{conv} \right) \cdot n_j^{phy} + \sum_{t \in \mathcal{T}^{dem}} \delta_t^{dem} \cdot d_t + \sum_{t \in \mathcal{L}} \delta_t^{oc} \cdot d_t^{oc}$$

$$= \sum_{t \in \mathcal{T}^{dem}} \delta_t^{dem} \cdot d_t + \sum_{t \in \mathcal{L}} \delta_t^{oc} \cdot d_t^{oc} + \sum_{j \in \mathcal{J}} \delta_j^{conv} \cdot n_j^{phy} + \sum_{j \in \mathcal{J}} \overline{c}_{j,k_j^*} \cdot n_j^{phy}$$

$$= z_u^{RMP} + \sum_{j \in \mathcal{J}} \overline{c}_{j,k_j^*} \cdot n_j^{phy}.$$

The first inequality derives from results above and the second inequality follows from the constraints (46), (47), and (48) in combination with $\overline{c}_{j,k_j^*} + \delta_j^{conv} \leq 0$. Similar bounds have been widely studied in the literature (e.g., see [108, 123] or [57]). Also another lower bound has been proposed by Farley [67] which is tight too. Vanderbeck [115] use a similar idea. Both, Farley [67] and Vanderbeck [115] replace the convexity constraint by lower and upper bounds on the sum of λs. The bound is obtained by dualizing all but the bound constraints.

The obtained lower bound can be used to terminate CGA since in practice solving the linear programming relaxation might be cumbersome because many iterations within CGA are necessary to proof optimality. This phenomenon is called *tailing-off effect*. Recall that we are solving an IP and thus tailing-off might occur at each node in the search tree.

Again, the lower bounds are just valid when the subproblems are solved to optimality. For instance when a partial or heuristic pricing is used, then optimal solutions to the j pricing subproblems denoted by k_j^* and hence LB_u are not available. But if partial pricing is performed, say the first promising column is picked, a valid lower bound is available at the end of CGA when a node has been explored.

Given the above mentioned lower bounds and the best upper bound UB^{IP} which is the best feasible solution found so far we can derive the following stopping criteria. At any given node in in the search tree, CGA is halted when the linear relaxation of the node problem is known. In Sect. 4.3.2 we use RMP optimality, i.e. $\overline{c}_j \geq 0$ $\forall j \in \mathcal{J}$. However, since we now have a lower bound on the optimal RMP solution LB_u at each iteration u of CGA, we stop solving the node solution if the optimality gap is closed, that is $z_u^{RMP} \geq LB_u$. Note that both stopping criteria are equivalent if the best lower bound LB_u, and hence LB_n, is obtained when CGA terminates. In some cases when degeneracy occurs the latter might hold whereas the former does not.

Moreover, we can also stop investigating the trajoctery when the computed best lower bound is worse than the best feasible solution, UB^{IP}, found so far in the course of the algorithm. The stopping criterion is $LB_u \geq UB^{IP}$ and if it holds we prune the node and backtrack to another live node which has not been investigated up to now. As mentioned above, if the optimal solution to the underlying problem is general integer, we can improve the lower bound as follows $LB_u \leftarrow \lceil LB_u \rceil$.

Using the lower bounds now available one can derive an upper bound cut on the subproblem solution. That might be helpful in proving that a node is infeasible and thus can be pruned in the search tree (e.g., see [76, 115]).

4.3.8.2 Aggregation of Subproblems

Models (45)–(49) and (52)–(78) aggregate all physicians form type j into one subproblem rather then using for each physician a separate subproblem since the working rules for type j physicians are identical. This approach implicitly takes into account subproblem symmetry. Considering two type j groups of physicians that have the same working rules and differ only in the value of, say, the preference parameters T_j^{win}. Recall, that this preference parameter is specified by each physician and denotes the length of the implicit starting time window. For instance, assume we have two groups of full-time physicians indexed by j_1 and j_2 but one has a 3-h and the other has a 1-h starting time window. In our original formulation these two types are treated separately as different subproblems. The fact that SP_{j_1} with $T_{j_1}^{win} = 1$ might generate a solution to SP_{j_2} with $T_{j_2}^{win} = 3$ argues for combining both in a single subproblem. The advantages are a reduction of computation time since one subproblem is included in another and an elimination of symmetry in the solution space but with the drawback of more complex single SP's. In what follows, we show the aggregation idea in absolute terms using the parameter T_j^{win}. However, the idea can be generalized to other parameters.

To combine two or more physician types j with the same working rules but different starting time windows needs some modification to the master problem and the corresponding subproblem(s). The new subproblem will use the least restrictive starting time preference which is 3 in the previous example. To guarantee that rosters with more restrictive time windows are generated by the solution process as well we have to extend constraints (48) in RMP by one covering constraint for each type j that is more restrictive than the least restrictive one. To state the new constraints in RMP and the new single subproblem j some notation is inevitable. Let $\mathcal{P}(j)$ be the set of physician type j with same working rules and different starting time windows except physicians with the least restrictive time window. The elements in $\mathcal{P}(j)$ are sorted in ascending order starting by the most restrictive. Let $T_{j,p}^{win}$ be the upper bound on the starting time window for each $p \in \mathcal{P}(j)$ and let T_j^{win} be the

4.3 Column Generation and B&P Algorithm

least restrictive time window. The set $\mathcal{K}_p(j)$ denotes the subset of schedules k for type j that fulfills the time window restriction with a predefined parameter $T_{j,p}^{win}$ for all $p \in \mathcal{P}(j)$. $\mathcal{K}(j)$ contains all schedules for aggregated physician type j. Finally, the number of physicians whose starting time window corresponds to $p \in \mathcal{P}(j)$ is designated by $n_{j,p}^{phy}$ and the total number of physicians aggregated in subproblem j is denoted by n_j^{phy}.

Using the previous notation we can state the new constrains that assure fulfillment of time window requirements for aggregated type j in RMP.

$$\sum_{k \in \mathcal{K}_p(j)} \lambda_{j,k} \geq \sum_{\rho=1}^{p} n_{j,p}^{phy}, \forall j \in \mathcal{J}, p \in \mathcal{P}(j) \tag{98}$$

For example, consider three types of full-time physicians FT_1, FT_2, FT_3 with 1-h, 2-h, and 3-h starting time window. The number of available physicians for type FT_1 is 2, for FT_2 is 3, and for FT_3 is 5. Now, we aggregate the three into one type named *aggregated full-time physician* (AFT). The set $\mathcal{P}(AFT)$ is defined as $\{FT_1, FT_2\}$ since FT_3 corresponds to the least restrictive time window. When $p = FT_1$, constraint (98) forces that at least $n_{AFT,FT_1}^{phy} = 2$ rosters that satisfy $T_{AFT,FT_1}^{win} = 1$ are included in a solution to RMP. When $p = FT_2$, constraint (98) ensures that at least 2 rosters with $T_{AFT,FT_1}^{win} = 1$ and 3 rosters with $T_{AFT,FT_2}^{win} = 2$ are assigned in a solution. The original convexity constraint (48) with modified parameters assures that no more physicians can be assigned than available physicians in aggregated group j which is $n_{AFT}^{phy} = 10$.

Since new constraints in RMP supply new dual variables the aggregated subproblem SP_j must be modified. Let $\delta_{j,p}^{conv}, \forall j \in \mathcal{J}, p \in \mathcal{P}(j)$, be the nonnegative dual values of constraints (98). To account for generating rosters with more restrictive time windows we introduce the binary variable $y_{j,p}^{win}$ and define it as follows

$$y_{j,p}^{win} = \begin{cases} 1, & \text{if time window corresponding to } p \in \mathcal{P}(j) \text{ is active} \\ 0, & \text{otherwise.} \end{cases}$$

The following objective function then replaces (52)

$$\text{Min} \sum_{w \in \mathcal{W}} \left(c_j^{paid} \cdot h_{j,w} + c_j^{over} \cdot o_{j,w} \right) - \sum_{t \in \mathcal{T}^{dem}} \delta_t^{dem} \cdot \left(x_{j,t} - y_{j,t}^{break} \right)$$
$$- \sum_{t \in \mathcal{L}} \delta_t^{oc} \cdot y_{j,t}^{oc} - \sum_{p \in \mathcal{P}(j)} \delta_{j,p}^{win} \cdot y_{j,p}^{win} - \delta_j^{conv} \tag{99}$$

Taking up our previous example the new objective function determines implicitly the time window that is used when generating a new roster with the aggregated subproblem. For instance, if we wish to generate a roster with $T_{AFT}^{win} = 3$ corresponding to the least restrictive case only δ_{AFT}^{conv} is relevant. On the other extreme, if we wish to generate a roster with the most restrictive time window, $T^{win}_{AFT,FT_1} = 1$, all $y^{win}_{AFT,p}$ with $p \in \mathcal{P}(AFT)$ are equal to 1 and all dual values $\delta^{conv}_{AFT,p}$ $p \in \mathcal{P}(AFT)$ and δ^{conv}_{AFT} are relevant. Since we want the aggregated subproblem to decide which case is priced out we have to introduce the following new constraints to the aggregated subproblem.

$$l_{j,w} - e_{j,w} \leq T^{win}_{j,p} \cdot y^{win}_{j,p}, \forall w \in \mathcal{W}, p \in \mathcal{P}(j) \tag{100}$$

$$l_{j,w} - e_{j,w} \geq \left(T^{win}_{j,p} + 1\right) \cdot \left(1 - y^{win}_{j,p}\right), \forall w \in \mathcal{W}, p \in \mathcal{P}(j) \tag{101}$$

Constraints (100), when tight, corresponding to $y^{win}_{j,p} = 1$, impose a time window restriction that is more restrictive than the least restrictive case to the aggregated subproblem. Constraints (101) force that, if $y^{win}_{j,p} = 0$, the generated rosters have at least a time window of $\left(T^{win}_{j,p} + 1\right)$. If constraints (100) are active then constraints (101) are not active and vice versa.

Again, using our example, if $y^{win}_{AFT,FT_1} = 1$, meaning that the subproblem prices out the roster with the most restrictive time window $T^{win}_{AFT,FT_1} = 1$, then it has also to be true that $y^{win}_{AFT,FT_2} = 1$. But when the final roster fulfills $T^{win}_{AFT,FT_2} = 2$ then y^{win}_{AFT,FT_1} has to be 0, otherwise subproblem SP_{AFT} would be infeasible. As mentioned, the just presented idea can be applied to other parameters or parameter combinations as long as it is true that the solution for the more restrictive SP is also feasible for the less restrictive.

Chapter 5
Experimental Investigations

In the following chapter after presenting the data set in Sect. 5.1 that is used in the computations we present three computational studies for the flexible shift scheduling problem of physicians in a hospital. First, in Sect. 5.2 we show results obtained by the heuristic decomposition approach (see Sect. 4.2) where 2-week schedules are the focus of interest. Among other, parameter analysis give insights into the underlying problem structure. In the second study in Sect. 5.3 we evaluate the B&P approach presented in Sect. 4.3. The aim is to find 2-, 4-, and 6-week schedules using real data from a anesthesia department. Finally, in the last study in Sect. 5.4 we solve all instances from the parameter analysis (see Sect. 5.2.2) again since as one can see we were not able to improve on the lower bounds. We use MPVarB to solve the instances to optimality. All algorithms are coded in Java and linked with CPLEX 10.2. We use the default setting of 0.01% for the optimality gap. All computations were performed on a 2 GHz PC (Intel Core2 CPU T7200) with 2.046 GB RAM running under the Windows VISTA operating system.

5.1 Input Data From MRI

In this section we summarize the data used in our computational study. We are using real data obtained by the anesthesia department of MRI which has approximately 1,100 beds available. In Sect. 5.1.1 we give some insight from the data analysis and present the different demand profiles that are used in the computations. In all experimental investigations the planning horizon is divided in periods of 1-h increments. The input demand profiles are based on realizations in different operating theaters at MRI in 2005 but the main scheduling object is the *central operating theater* (ZOP). Furthermore, we introduce a basic setting of parameters in Sect. 5.1.2 that will be used in the computations and has been used in the application presented in Sect. 3.3. For instance, we have to specify the minimum and maximum shift length as well as the minimum rest length between two consecutive shifts and/or on-call services.

J.O. Brunner, *Flexible Shift Planning in the Service Industry*, Lecture Notes
in Economics and Mathematical Systems 640, DOI 10.1007/978-3-642-10517-3_5,
© Springer-Verlag Berlin Heidelberg 2010

5.1.1 Demand Profiles

Our cooperation partner MRI has 14 distinct operating theaters each has between one up to 8 rooms where surgeries can be performed. Furthermore, the workforce in 2005 was approximately 60 anesthetists including full- and part-timers. Beside conducting operations in the theaters the anesthesia department has a pre-medication ambulance which has the task to wise up and check the patients before an operation is carried out, and is responsible for the emergency doctor. Additionally, among various other tasks the anesthetists has to do research and has to conduct a duty that constitutes of visiting all departments in the hospital to treat patients that have ache.

MRI provided us with the data from the operating theaters. The data included the time of the first and last contact before and after an operation with the patient. This time span is used to determine the demands in each operating theater. We have discretized the demand per period which might be half an hour or 1 h long, by counting the number of anesthetists working actively. Figure 5.1 shows the average demand for operations on a normal working day that is Monday through Friday per each half hour period. The x-axis denotes the time within the day whereas the y-axis characterizes the number of needed anesthetists. As can be seen the first operations start in the interval from 7:30 to 8 a.m., then the number of necessary anesthetists rises up to 29 on average until approximately 10 a.m., and finally decreases to 0 until 9 p.m. In our data analysis we aggregated the demand over all operating theaters and considered just elective surgeries that were planned some time in advance. Elective surgeries are more than 85% of all surgeries performed at MRI. Sure, there has to be some more staff to cope with emergencies but in off hours the hospital has on-call services for that purpose and during normal working hours the demand might be increased virtually on experience. Note that the discretization overestimates the requirements per period. For example, an anesthetist is in contact with the first patient until 10:15 a.m. and takes over the second patient at 10:40 a.m. then if we apply 1 h periods in the discretization the identical anesthetist is counted twice.

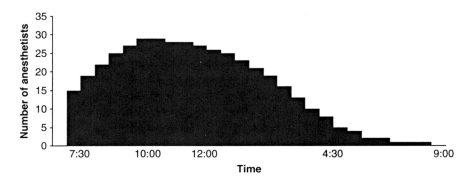

Fig. 5.1 Average demand per half hour period in 2005

5.1 Input Data From MRI

Table 5.1 Average day demand on Monday through Friday in 2005

Period	Mo	Tu	We	Th	Fr	$MEAN$	VAR
0:00	0	0	0	0	0	0	0
1:00	0	0	0	0	0	0	0
2:00	0	0	0	0	0	0	0
3:00	0	0	0	0	0	0	0
4:00	0	0	0	0	0	0	0
5:00	0	0	0	0	0	0	0
6:00	0	0	0	0	0	0	0
7:00	18	20	19	2	17	15	55.7
8:00	21	24	22	20	21	22	2.3
9:00	27	32	30	25	27	28	7.7
10:00	31	34	33	31	29	32	3.8
11:00	30	34	33	32	29	32	4.3
12:00	29	33	31	32	26	30	7.7
13:00	27	29	29	29	21	27	12
14:00	22	25	24	24	17	22	10.3
15:00	17	18	19	19	10	17	14.3
16:00	11	11	12	12	6	10	6.3
17:00	6	6	6	6	3	5	1.8
18:00	3	3	3	3	1	3	0.8
19:00	1	1	1	1	0	1	0.2
20:00	0	1	1	1	0	1	0.3
21:00	0	0	0	0	0	0	0
22:00	0	0	0	0	0	0	0
23:00	0	0	0	0	0	0	0

The overestimation is smaller when the periods are shorter. Consider the case of half hour long periods in the previous example.

Table 5.1 displays the demand per 1 h period for the average workweek Monday–Friday in 2005. The first column denotes the period where the bold times signal p.m. hours. Column 2–6 give the average demand on the specific day of the week. The next column give the mean demand per 1 h period. Eventually, the last column gives the variance over the average demand per day. The demand does not vary just form one period to the next within a day but also fluctuates from one day to the next in the week. For instance, the maximal average demand on Tuesday in 2005 is 34 whereas the maximal average demand on Friday in 2005 is 29. Variances are high at the beginning of the day and after noon. Figure 5.2 shows the fluctuating demand for anesthetists in 2005 based on a 1-h discretization aggregated over all operating rooms.

From our data analysis we conclude that:

- Regular demand periods (1 h) span from 7 a.m. to 9 p.m.,
- Demand for service follows multi-seasonality effects from period to period.
- Demand fluctuates highly from one operating theater to the next.

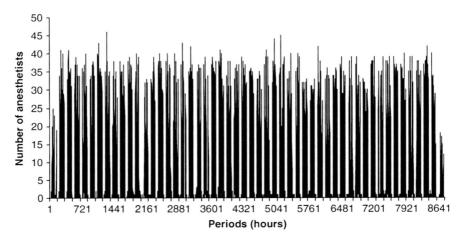

Fig. 5.2 Hospital-wide cumulated demand in 2005

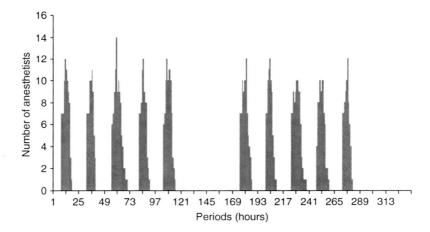

Fig. 5.3 2-week demand in ZOP

- No shift can cover all (average) demand periods within a day.
- The current scheduling procedure (see Sect. 3.3) is inadequate.

Based on the insights we have decided to consider the requirements in ZOP since the number of operations carried out are more or less constant over the year. ZOP is the biggest operating theater at MRI and has 8 operating rooms where surgeries can be performed. So, the basic demand profile that has been used in the example is depicted in Figure 5.3. Furthermore, it corresponds to the dark area and reflects requirements in ZOP at MRI for 2 weeks in January 2005. The x-axis references the periods in hourly increments and the y-axis gives the demand, which varies widely from period to period and from 1 day to the next. In addition, some anesthetist are

5.1 Input Data From MRI

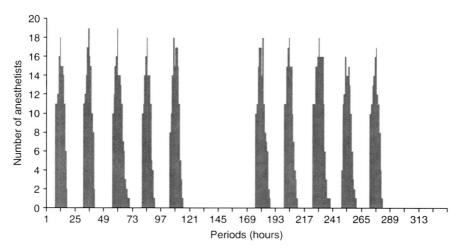

Fig. 5.4 Aggregated 2-week demand in ZOP, HNOP, and SPOP (profile 2)

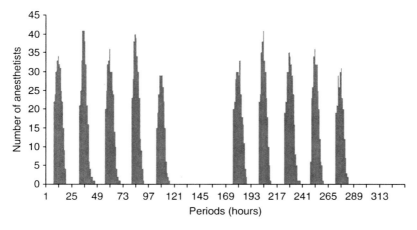

Fig. 5.5 Aggregated 2-week demand in all operating rooms (profile 3)

needed for on-call services on each day in the planning horizon (i.e., $d_t^{oc} > 0$, $\forall t \in \mathcal{L}$). The number of anesthetists necessary to cover on-call demand may vary from 1 day to the next. Regular working days are Monday through Friday. On Saturday and Sunday only on-call service is required. The on-call service regulations as presented in Sect. 3.2.1 are in use. In our basic computations we assume the on-call demand to be constant over the planning horizon and in the simplest version we set $d_t^{oc} = 1$, $\forall t \in \mathcal{L}$ for operating theaters like ZOP.

Two further demand profiles based on data obtained from the anesthesia department at MRI for 2 weeks in January 2005 are presented in the following. Profile 2 (cf. Fig. 5.4) reflects the cumulative demand in ZOP, the *ear, nose and throat operating room* (HNOP), and the *sport operating room* (SPOP). Profile 3 (cf. Fig. 5.5)

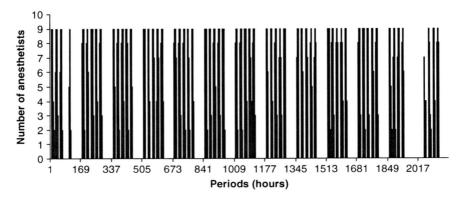

Fig. 5.6 Demand profile central operating theater first quarter in 2005

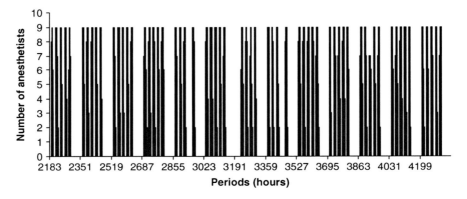

Fig. 5.7 Demand profile central operating theater second quarter in 2005

considers the hospital-wide demand for anesthetists which includes a total of 14 distinguish locations where surgeries are performed. All input data is collected for the same 2-week period in January 2005. Profile 1 is for the ZOP only and is the demand profile shown in Fig. 5.3.

For the second study presented in Sect. 5.3 we use the complete data available in 2005 for ZOP. The data is depicted in Figs. 5.6–5.9. To count for the error we make by discreticizing we cut the demand that is higher then 8 anesthetists off. Recall, ZOP consists of 8 operating rooms. Subsequently, we increment each period with positive demand by 1 because one anesthetist is needed in the wake up room. Each of the four Figs. 5.6–5.9 shows the demand for one quarter in 2005.

5.1 Input Data From MRI

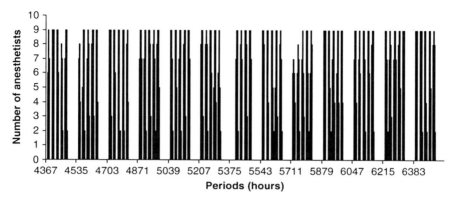

Fig. 5.8 Demand profile central operating theater third quarter in 2005

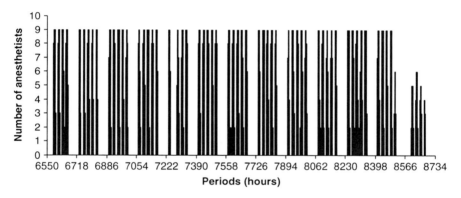

Fig. 5.9 Demand profile central operating theater fourth quarter in 2005

5.1.2 Basic Parameter Settings

The objective of the second part in this section is to determine the minimum number of anesthetists needed to satisfy demand during the 336 periods (cf. Fig. 5.3). We must consider the peak demand of 14 on Wednesday in the first week ($t = 60$ in Fig. 5.3), as well as the implications of on-call services. Because an anesthetist cannot be on duty on the day following an on-call service and due to the break assignment for the fixed shift within an on-call service, we need a total of 16 identical anesthetists, i.e. we have 16 full-time anesthetists. In other words, we do not consider different groups (i.e. $|\mathcal{J}| = 1$).

Furthermore, we assume in our basic parameter settings that a shift must span at least 7 consecutive hours, and in accordance with the general labor contract (see [88]), not exceeds 13 h. In addition, each shift must be given an hour-long break that can be assigned anywhere between a 3-h pre- and a 3-h post-break workstretch. A maximum pre-break workstretch is not applied in the basic settings.

Table 5.2 Basic parameter settings

$\lvert\mathcal{J}\rvert$	=	1	T_1^{win}	=	3	T^{oc}	=	24	$\overline{o}_{1,i}$	=	6	$\forall i \in \mathcal{I}_1$

$\lvert\mathcal{J}\rvert$ = 1	T_1^{win} = 3	T^{oc} = 24	$\overline{o}_{1,i}$ = 6	$\forall i \in \mathcal{I}_1$
$\lvert\mathcal{I}_1\rvert$ = 16	$\underline{T}_1^{shift}$ = 7	$\overline{n}^{oc}$ = 1	$r_{1,i}$ = 42	$\forall i \in \mathcal{I}_1$
$\lvert T\rvert$ = 336	$\overline{T}_1^{shift}$ = 13	$\overline{T}_1^{pre}$ = 9	d_t^{oc} = 1	$\forall t \in \mathcal{L}$
$\lvert\mathcal{W}\rvert$ = 2	$\underline{T}_1^{rest}$ = 12	$\underline{T}^{post}$ = 3	$\underline{T}_1^{pre}$ = 3	
$\lvert\mathcal{L}\rvert$ = 14				

Each anesthetist has a contract that specifies 42 regular hours per week; i.e., $r_{j,i} = 42, \forall j \in \mathcal{J}, i \in \mathcal{I}_j$, but can be scheduled for up to six additional hours of overtime, i.e., $\overline{o}_{j,i} = 6$. The number of on-call services, $\overline{n}_j^{oc}$, that an anesthetist is permitted to be assigned in a week is 1. Finally, we allow a 3-h time window for starting different shifts within a week, i.e. $T_j^{win} = 3 \,\forall j \in \mathcal{J}$. Table 5.2 summarizes the basic parameter values.

As previously stated, our aim is to cover demand with regular hours first, then with overtime, and finally with outside hours when the first two options are exhausted. To ensure the desired order is maintained, we apply the logic below for the three options using relative costs instead of absolute costs to facilitate the presentation. This is permissible because the anesthetists are assumed to be identical. Given that an anesthetist i associated with group j can be assigned no more than $\overline{o}_{j,i} = 6\,\text{h}$ of overtime per week, the cost parameters should satisfy the condition $c^{out} > \sum_{j \in \mathcal{J}} \sum_{i \in \mathcal{I}_j} \overline{o}_{j,i} \cdot c_{j,i}^{over}$ and $c_{j,i}^{over} > c_{j,i}^{paid}$. With this in mind, we set $c_{1,i}^{paid} = 0.1$, $c_{1,i}^{over} = 1$, and $c^{out} = 100$ for the computations. The cost to cover an on-call service by an outside physician is 1,000, i.e. $c^{oc\_out} = 1,000$.

5.2 Heuristic Decomposition

In this section, we begin by investigating how the requirement for break assignments and the presence of time windows affect solution quality. This is followed by a parametric study of shift lengths, time window lengths, and number of physicians. The analysis is structured around the example given in Sect. 3.3. Finally, we examine several variations of real-world instances to determine how well CPLEX does in finding optimal solutions. In all cases, the preprocessing procedure and heuristic decomposition strategy described in Sects. 4.1 and 4.2 are used. We consider just one group of full-time physicians. Although our aim is to construct 2-week schedules, it is straightforward to extend the analysis to any number of weeks since the decomposition heuristic allows us to derive schedules for each week separately.

Throughout this section, we use the same demand profile (see Fig. 3.5) introduced in Sect. 3.3, break assignment procedure, and parameter setup employed for the case study in Sect. 3.3. We refer to those results as the *base case*. Each instance in Sects. 5.2.1 and 5.2.2 has been solved twice, first with our decomposition heuristic with a time limit of 3,600 s per 1-week model, and then as a full 2-week MIP

5.2 Heuristic Decomposition 81

Table 5.3 Problem size for weekly model

No. anesthetists (on-call)	Total no. variables	No. binary variables	No. SOS constraints	No. rows
16 (1)	7,067	5,488	80	11,457

Table 5.4 Comparison of model features

		Decomposition								2-week MIP	
Breaks	Time win-dows	Total rel. costs	Week	Run-time (s)	Opt. gap (%)	Over-time hours	Under-time hours	Idle-time hours	No. out-side	Total rel. costs	Run-time (s)
Yes	Yes	0.0^a	1	163	0.0	0	62	155	0	0.0^a	3,999
			2	109	0.0	0	92	133	0		
Yes	No	0.0^a	1	95	0.0	0	57	160	0	0.0^a	2,349
			2	130	0.0	0	72	153	0		
No	Yes	0.0^a	1	43	0.0	0	112	105	0	0.0^a	524
			2	37	0.0	0	134	91	0		
No	No	0.0^a	1	37	0.0	0	117	100	0	0.0^a	281
			2	50	0.0	0	126	99	0		

[a] Optimal objective function value

with a time limit of 7,200 s (see Chap. 3). Whenever we are unable to achieve convergence with one of the two methods we employ the problem-specific branch-and-price algorithm introduced in Sect. 4.3 to compute the optimal solution. The results obtained by the branch-and-price algorithm are discussed in more detail in Sect. 5.4. In Sect. 5.2.3 we employ only the heuristic decomposition to generate the 2-week schedules.

Table 5.3 gives the number of variables and constraints in a 1-week instance of the problem after preprocessing when we assume that break placement and time window constraints are active. Furthermore, the demand profile for ZOP depicted in Fig. 5.3 is used to calculate the values in Table 5.3.

Comparisons are based on objective function values using relative cost coefficients. Of the 16 instances investigated, we have found that no overtime is necessary and hence no paid out hours are required. This result is due to the large difference between supply and demand, an artifact of the inefficient staffing levels and procedures currently employed. Tables 5.4–5.9 report the output statistics for the 16 runs. As can be seen, the decomposition heuristic obtains equal or better objective function values than the 2-week MIP in the same or considerably less time, and in 12 instances, finds the optimum.

To a large extent, undertime is a indication of how closely a schedule matches the demand profile. Recalling the settings in Sect. 3.3, there are 455 h of demand in the first week of the 2 week problem considered and 447 h in the second. The supply in both weeks is 672 h, assuming 16 anesthetists each working 42 regular hours per week. For example, if the undertime in week one is 217 h (= 672–455 h) and the objective function is 0, then a perfect match has been achieved and hence

Table 5.5 Computational results for parameter $\overline{T}_j^{shift}$

Max shift length	Total rel. costs	Week	Run-time (s)	Opt. gap (%)	Over-time hours	Under-time hours	Idle-time hours	No. out-side	Total rel. costs	Run-time (s)
					Decomposition				2-week MIP	
13	0.0[b]	1	163	0.0	0	62	155	0	0.0[b]	3,999
		2	109	0.0	0	92	133	0		
12	0.0[b]	1	170	0.0	0	78	139	0	0.0[b]	1,641
		2	377	0.0	0	97	128	0		
11	0.0[b]	1	113	0.0	0	85	132	0	100.0	7,200
		2	101	0.0	0	110	115	0		
10	100.0[b]	1	3,600	$-$[a]	0	105	112	1	400.0	7,200
		2	86	0.0	0	104	121	0		
9	100.0[b]	1	3,600	$-$[a]	0	126	91	1	200.0	7,200
		2	85	0.0	0	122	103	0		

[a] Optimum not found and no improvement in lower bound
[b] Optimal objective function value

Table 5.6 Computational results for parameter T_j^{win}

Time window (h)	Total rel. costs	Week	Run-time (s)	Opt. gap (%)	Over-time hours	Under-time hours	Idle-time hours	No. out-side	Total rel. costs	Run-time (s)
					Decomposition				2-week MIP	
0	1300.0[c]	1	3,600	$-$[a]	0	104	113	9	$-$	7,200
		2	3,600	$-$[a]	0	136	89	4		
1	100.0[c]	1	160	0.0	0	63	154	0	$-$	7,200
		2	3,600	$-$[a]	0	93	132	1		
2	0.0[b]	1	164	0.0	0	64	153	0	$-$	7,200
		2	126	0.0	0	85	140	0		
3	0.0[b]	1	163	0.0	0	62	155	0	0.0[b]	3,999
		2	109	0.0	0	92	133	0		

[a] Optimum not found and no improvement in lower bound
[b] Optimal objective function value
[c] Optimal objective function value is 0.0

the final rosters do not have idletime. No outside resources or overtime are needed. As a rule, if two schedules have the same objective function value, then the one with the greater amount of undertime would be preferred. This allows supervisors more flexibility in meeting the requests and preferences of the individual physicians.

5.2.1 Analysis of Different Model Features

To assess the impact of break considerations and time window requirements on solution quality, we compared the base case, which includes both of these options,

5.2 Heuristic Decomposition

Table 5.7 Computational results for number of physicians

No. anesth-etists	Total supply (h)	Decomposition								2-week MIP	
		Total rel. costs	Week	Run-time (s)	Opt. gap (%)	Over-time hours	Under-time hours	Idle-time hours	No. out-side	Total rel. costs	Run-time (s)
16	672	0.0[b]	1	163	0.0	0	62	155	0	0.0[b]	3,999
			2	109	0.0	0	92	133	0		
15	630	100.0[b]	1	3,600	−[a]	0	43	132	1	200.0	7,200
			2	157	0.0	0	38	145	0		
14	588	300.0[c]	1	3,600	50.0	0	29	104	2	300.0	7,200
			2	3,600	−[a]	0	24	117	1		
13	546	400.0[b]	1	390	0.0	0	9	82	3	400.0[b]	6,857
			2	3,600	−[a]	0	9	90	1		

[a] Optimum not found and no improvement in lower bound.
[b] Optimal objective function value.
[c] Optimal objective function value is 200.0.

Table 5.8 Problem size for weekly model for different demand profiles

Profile no.	No. anesthetists (on-call)	Total no. variables	No. binary variables	No. integer variables	No. SOS sets	No. rows
1	16 (1)	8,153	2,816	168	80	15,903
2	23 (2)	11,646	4,048	168	115	22,784
3	47 (3)	23,622	8,272	168	235	46,376

Table 5.9 Computational results for different demand profiles

Profile no.	Total rel. costs	Week	Total demand (h)	Total supply (h)	Run-time (s)	Optimality gap (%)	Over-time hours	Under-time hours	Idle-time hours	No. out-side
1	0.0	1	455	672	2,179	0.0	0	50	167	0
		2	447	672	929	0.0	0	74	151	0
2	200	1	725	966	18,000	−[a]	0	63	178	2
		2	749	966	3,576	0.0	0	78	139	0
3	0.0	1	1,464	1,974	12,514	0.0	0	131	379	0
		2	1,450	1,974	6,086	0.0	0	158	366	0

[a] Optimum not found and no improvement in lower bound.

with the three other scenarios denoted by the pairs (yes, no), (no, yes), (no, no). A second statistic of interest was the runtime.

Table 5.4 displays the computational results. The runtimes suggest that including one or both options in the model may increase the computation time significantly. Without breaks or time windows, runtimes with 37 and 50 s are reported, while in the most restricted case, these values increase up to 163 s in the first week and 109 s in the second. However, the final schedules are of higher quality with respect to the general labor rules and physician preferences. When only breaks are present (no time window constraints), CPLEX took approximately 4 min to converge. For the case where time window constraints are present but no break assignments, CPLEX found an optimal solution in slightly more than 1 min. Only 43 s are needed for the first week and 37 s for the second. The schedule includes 112 h of undertime in week 1 and 134 h in week 2, and no overtime or outside resources in either week. Compared to our base case where breaks of 1 h length are present, solution times decrease by 73.62 and 66.06% in each week, respectively. Moreover, fewer working hours are required to cover the demand, thus offering more flexibility in the use of undertime. From a computational point of view, it appears that including the break constraints increases the difficulty of a problem instance so it may be best to omit them from the model and follow the current procedure of assigning them as the day permits. By way of comparison, the 2-week MIP required considerably more time than the decomposition heuristic to solve the four instances.

5.2.2 Parametric Analysis

In this section, we focus on the base case and investigate how the solution changes when the settings for the maximum shift length $\overline{T}_j^{shift}$, the time windows T_j^{win}, and the number of physicians are varied. Also, the values given in Table 5.3 for the problem size remain the same unless stated otherwise.

5.2.2.1 Maximum Shift Length

The first investigation concerns the maximum shift length $\overline{T}_j^{shift}$. In the base case this value is set to 13. We now consider reducing it down to 9 in hourly increments while maintaining the 1-h break. The results are summarized in Table 5.5 for each of the 5 cases. In all instances no overtime is used so the relative costs are proportional to the real costs. If overtime is present, the paid out time plus the use of outside resources would determine the final costs for a schedule. Overtime in the second week could be then considered as paid out time or could serve as initial conditions for a third week if a rolling time horizon is considered. In several cases, no gap is reported because CPLEX was not able to improve on the LP solution of 0.0 at the root node. However, the decomposition heuristic does find the optimal objective function value for each instance as opposed to the 2-week MIP, which does not in three of the four new instances.

The first two rows in Table 5.5 correspond to the base case with $\overline{T}_j^{shift} = 13$. For the two cases where $\overline{T}_j^{shift} = 12$ or 11 CPLEX converges to an optimal solution with zero cost, and in the cases 10 and 9 CPLEX fails to converge within the 1-h time limit for all but one week. From this result, we conjecture that when the optimality gap can be closed, solution times are reasonable. This is due, in part, to the fact that there are a large number of alternative optima, which facilitate convergence. As the shift length shortens, it becomes more difficult to find an optimal combination of

5.2 Heuristic Decomposition

shifts that covers the demand without introducing unnecessary overtime. When the maximum shift length is 10, one outside anesthetist is required to cover the demand. As $\overline{T}_j^{shift}$ is reduced to 9, this requirement remains 1. From a managerial point of view, in cases like this, it may be possible to cover one or more of those outside hours without incurring additional costs by asking an physician on duty to work beyond his 9-h shift, assuming that there is undertime in his schedule. Restricting shifts lengths to 9 or fewer hours with the caveat that a physician (an anesthetist) may have to work an additional hour on occasion may be preferable to assigning 10–13 h shifts on a routine basis.

5.2.2.2 Time Window Length

We next investigate how the length of the time window, T_j^{win}, affects the solution. We have seen in Table 5.4 that when there are break assignments but no time window restrictions, it takes a long time to solve the problem to optimality. Preference considerations suggest that the time window should be as short as possible so we begin with a width of 0 and look at all cases in hourly increments up to 3, which is equivalent to 4 starting points. Table 5.6 summarizes the results for a maximum shift length of 13. When no daily variations in starting times are permitted ($T_j^{win} = 0$), the results indicate that a total of 13 outside hours are needed to cover the demand. The quality of the corresponding schedule is relatively low, given that it contains considerable undertime and the use of outside resources. As soon as we allow a 1-h time window, the relative costs drop from 1,300 to 100, and only 1 outside hour is required.

When $T_j^{win} = 2$, an optimal schedule with a total of 149 h of undertime is generated. Also, the base case with $T_j^{win} = 3$ gives an optimal schedule. From the results we see that as T_j^{win} increases convergence improves. When the time windows are tight, some amount of outside hours are required to cover demand. In such cases, CPLEX is not able to close the optimality gap because it could not find any cuts that increased the lower bound. Moreover, in three of the four instances, CPLEX fails to find a feasible solution to the 2-week MIP within the 2-h time limit.

5.2.2.3 Number of Physicians

The final set of runs discussed in this section is aimed at understanding the relationship between the number of full-time physicians and the shortfall in coverage. The tradeoff between the use of permanent staff and outside resources to cover demand is a critical management consideration when deciding how to structure the permanent workforce. In the analysis, the number of physicians is reduced incrementally from 16 in the base case down to 13. The size of the weekly scheduling problems decreases in proportion to the vector (no. variables, no. binary variables, no. SOS constraints, no. rows) = (438, 343, 5, 712). For example, for 15 physicians, the total number of variables is $7,067 - 438 = 6629$. Table 5.7 shows the results.

When the number of physicians is reduced below 16 it should be clear from the discussion in Sect. 5.1.2 that the number of outside hours must increase, since on the busiest day a minimum of 16 physicians are required to cover the demand. In all cases, no overtime is needed and consequently no hours are paid out. The last column in Table 5.7 shows that the number of outside hours increases by 1, then by 3, and finally by 4 for the three new scenarios. Optimality was achieved in four cases, and in a fifth, the gap was 50.0% for the first 1-week problem. In the remaining cases, the gap could not be determined for the 1-week problems because CPLEX failed to improve on the lower bound. Neverthelss, the decompostition strategy derives the optimal schedules in three out of the four instances for the 2-week problem whereas the 2-week MIP reports optimal schedules in only two cases.

The results dramatically point out the inefficiency of staffing practices and levels currently in use. By reducing the number of full-time physicians to 15, for example, only one additional outside hour is needed over the 2-week planning horizon. Even when only 13 physicians are available, 4 outside hours are all that are needed to cover demand. And in no case is overtime required to cover demand. Finally, we see that as the number of physicians is reduced, the amount of undertime decreases since there are fewer regular hours available. In all, this phase of the analysis shows that if the weekly demand over the year is reasonably steady, then significant savings can be realized by reducing the size of the full-time staff.

5.2.3 Analysis of Instances of Different Sites

In the last set of experiments, we compare the results obtained for the base case, which we now refer to as profile 1, with those obtained for two larger instances.

The size of the integer programs investigated in this section corresponding to demand profiles 2 and 3 that are presented in Figs. 5.4 and 5.5 is given in Table 5.8. The first column of the table lists the profile number. The remaining columns are the same as those in Table 5.3 and are self-explanatory.

The more physicians available, the larger the instance, but as we shall see, there is not a strong correlation between size and difficulty (Table 5.3). Nevertheless, we have raised the time limit for the new profiles to 18,000 s (5 h) for each CPLEX call. Table 5.9 summarizes the results. The entries are almost identical to those in the previous tables except that two new columns have been added to show the total demand and total available supply in each week. In all but one case (profile 2, week 1), the optimal solution is found, and for that case, only two outside hours are needed; no outside hours are needed otherwise. Also, in no case are overtime hours used and consequently no paid out hours. As mentioned previously, 161 h of overtime are used in practice for profile 1.

Although it is not possible to find the optimal solution for profile 2 within the 5-h time limit, the feasible schedule reported has a relatively low cost of 200, a total of

5.3 B&P Algorithm

141 h of undertime, and a need for two outside hours in week one. Compared with current practice that use 263 h of overtime but no outside resources (the 'unofficial' rule discussed in Sect. 3.3.2 requires three physicians to work longer than their 9-h shift), the new schedule is markedly superior, especially when the assignment of breaks, the adherence to work rules and the absence of overtime are taken into account.

For profile 3 with the hospital-wide demand, the two runs have yielded optimal solutions with a relative cost of 0.0. In the next to last column in Table 5.9 we see that the undertime in week one is 131 h and in week two 158 h. Compared to current practice, this is a substantial improvement and indirectly represents a savings of 485 h of overtime (the 'unofficial' rule requires five physicians to worker beyond their shifts). This result shows that high quality solutions can be obtained for almost 50 physicians within a bit more than 5 h – much less time than is now used to generate schedules manually.

5.3 B&P Algorithm

The proposed B&P algorithm is evaluated using data introduced in Sect. 5.1. In contrast to the previous investigations in Sect. 5.2 we extend the workforce by considering two groups of physicians ($|J| = 2$). The first consisting of full-timers (FTs) which are defined by constraints (53)–(78) and the second part-timers (PTs) which are defined by (53)–(59) and corresponding variable definitions. Furthermore, we generate schedules for up to 6-weeks in this section using both branching strategies MPVarB and SPVarB presented in Sects. 4.3.4 and 4.3.5, respectively. None of the problems investigated here could be solved with CPLEX within a 2-h time limit to optimality.

For the FT physicians we use almost the basic parameter settings (cf. Table 5.2), whereas for the PT physicians a modified set of parameters is used. Both sets are discussed in the following.

Table 5.10 summarizes the parameter values for an arbitrary FT. As can be seen, full-time physicians must be assigned shifts of at least 7 and at most 13 consecutive hours inclusive of an 1-h break. Once a shift starts, the break must occur within the first 6 h. After a shift ends there must be a rest period of at least 12 h before the next shift begins. Finally, an FT can work at most 1 on-call service per week, and each

Table 5.10 Parameter settings for arbitrary FT

$\overline{T}^{shift}_{FT} = 13$	$\underline{T}^{shift}_{FT} = 7$	$\overline{T}^{rest}_{FT} = 12$
$\underline{T}^{pre}_{FT} = 3$	$\overline{T}^{pre}_{FT} = 6$	$\underline{T}^{post}_{FT} = 3$
$T^{win}_{FT} = 3$	$n^{phy}_{FT} = 12$	$r_{FT} = 42$
$\overline{o}_{FT} = 6$	$\overline{n}^{oc}_{FT} = 1$	
$c^{over}_{FT} = 10$	$c^{paid}_{FT} = 1$	

Table 5.11 Parameter settings for arbitrary PT

$\overline{T}_{PT}^{shift} = 6$	$\underline{T}_{PT}^{shift} = 4$	$\overline{T}_{PT}^{rest} = 12$
$n_{PT}^{phy} = 12$	$r_{PT} = 21$	$\overline{o}_{PT} = 3$
$c_{PT}^{over} = 10$	$c_{PT}^{paid} = 1$	

shift within a week must start within a 3-period time window. All hourly costs are normalized and scaled appropriately.

The parameter settings for an arbitrary PT are given in Table 5.10. By way of contrast, a PT works between 4 and 6 h when on duty and has no starting times restrictions from day to day. As a whole, this group has not the skill to work on-call services and by design is not given a break during a shift.

The cost, c^{out}, to hire an outside physician for one period $t \in \mathcal{T}^{dem}$ is set to 1,000 whereas the cost, c^{oc_out}, to hire an outside physician to conduct an on-call service starting in period $t \in \mathcal{L}$ is set to 10,000. Again, we use relative cost values instead of absolute values to facilitate the presentation. The sets $\mathcal{T}$, $\mathcal{L}$, and $\mathcal{W}$ depend on the problem instance considered and can be inferred from the input data given in Tables 5.11–5.14. For all tests we set $n_{FT}^{phy} = 12$, $n_{PT}^{phy} = 6$, and the length of an on-call service to $T^{oc} = 24$ h. The demand profiles depicted in Figs. 5.6–5.9 are used in the computations. The on-call demand is set to one which is $d_t^{oc} = 1$ for all $t \in \mathcal{L}$. Figure A.2 defines the function f_1 in constraints (58) which calculates the number of hours that are charged for any on-call service starting in period $t \in \mathcal{L}$, and Fig. A.1 shows the holiday profile in 2005. Both figures are given in the Appendix.

For the computations in this section we set working memory to 254 MB. The default settings of CPLEX are used when solving RMP but when the SPs are solved we set the frequency parameter for calling the internal heuristic to find integer solutions at a node to 100. Moreover, each SP is solved from scratch with the MIP emphasis parameter set to *feasibility*. The computations are halted when a promising column is encountered, rather than when optimality is confirmed. When our feasibility heuristic is called to solve the RMP as an IP, the default setting of 0.01% for the optimality gap is used.

For each problem instance, the B&P algorithm was run twice, first using MPVarB and then using SPVarB. In all cases, the supply and demand are such that the optimal objective function values were 0. This allowed us to better assess the solution quality. Termination occurs when either the optimal solution is found or when the search tree reaches a preset limit of 100 nodes. In the following sections, we present results for 2-, 4-, and 6-week problem instances. Performance is measured by objective function values and runtimes. We conclude the second study by drawing general observations in Sect. 5.3.4.

Table 5.12 Results for 2-week problems

Prob no.	Weeks	Total dem hours	Total sup hours	Time [s] start heur	Ini LP soln	Ini IP soln	MPVarB Time [s] algo	MPVarB Final IP soln	MPVarB no. Nodes	MPVarB no. Cols gen	SPVarB Time [s] algo	SPVarB Final IP soln	SPVarB no. Nodes	SPVarB no. Cols gen
1	1–2	760	1140	4.847	0	0	0.135	0	1	46	0.130	0	1	46
2	3–4	929	1260	3.148	0	0	0.577	0	1	53	0.581	0	1	53
3	5–6	924	1260	15.812	0	0	0.195	0	1	47	0.193	0	1	47
4	7–8	934	1260	6.657	0	0	0.553	0	1	52	0.566	0	1	52
5	9–10	857	1260	3.243	0	0	0.185	0	1	47	0.177	0	1	47
6	11–12	802	1140	18.761	0	0	0.181	0	1	48	0.176	0	1	48
7	13–14	724	1140	2.437	0	0	0.240	0	1	47	0.238	0	1	47
8	15–16	873	1260	5.844	0	0	0.183	0	1	47	0.168	0	1	47
9	17–18	829	1140	3.597	0	0	0.274	0	1	50	0.265	0	1	50
10	19–20	817	1140	16.784	0	0	0.204	0	1	48	0.203	0	1	48
11	21–22	814	1140	14.889	0	0	0.173	0	1	48	0.172	0	1	48
12	23–24	883	1260	4.205	0	0	0.191	0	1	48	0.191	0	1	48
14	27–28	796	1260	4.585	0	0	0.093	0	1	46	0.105	0	1	46
15	29–30	895	1260	4.557	0	0	0.467	0	1	48	0.469	0	1	48
16	31–32	893	1260	3.733	0	0	0.180	0	1	48	0.176	0	1	48
17	33–34	789	1140	2.805	0	0	0.197	0	1	47	0.183	0	1	47
18	35–36	833	1260	11.434	0	0	0.230	0	1	48	0.226	0	1	48
19	37–38	941	1260	22.676	0	0	0.407	0	1	50	0.401	0	1	50
20	39–40	843	1140	16.023	0	0	0.203	0	1	48	0.200	0	1	48
21	41–42	864	1260	11.979	0	0	0.264	0	1	47	0.259	0	1	47
22	43–44	786	1140	5.964	0	0	0.222	0	1	49	0.217	0	1	49
23	45–46	927	1260	6.768	0	0	0.222	0	1	47	0.225	0	1	47
24	47–48	885	1260	27.537	0	0	0.203	0	1	47	0.199	0	1	47
25	49–50	908	1260	7.780	0	0	0.415	0	1	52	0.408	0	1	52
26	51–52	566	1140	12.582	0	0	0.077	0	1	45	0.077	0	1	45

Table 5.13 Results for 4-week problems

Prob no.	Weeks	Total dem hours	Total sup hours	Time (s) start heur	Ini LP soln	Ini IP soln	MPVarB Time (s) algo	Final IP soln	no. Nodes	no. Cols gen	SPVarB Time (s) algo	Final IP soln	no. Nodes	no. Cols gen
27	1–4	1697	2400	59.755	0	0	0.939	0	1	68	0.922	0	1	68
28	5–8	1866	2520	59.767	0	0	2.693	0	1	78	2.702	0	1	78
29	9–12	1667	2400	82.853	0	0	0.714	0	1	64	0.448	0	1	64
30	13–16	1605	2400	49.278	0	1000	3.275	0	11	74	4.373	0	11	78
31	17–20	1654	2280	62.384	0	0	1.975	0	1	72	1.974	0	1	72
32	21–24	1705	2400	66.880	0	0	0.384	0	1	64	0.399	0	1	64
33	25–28	1701	2520	62.952	0	0	0.605	0	1	65	0.647	0	1	65
34	29–32	1796	2520	74.352	0	1000	2.204	0	9	68	2.227	0	6	68
35	33–36	1630	2400	58.346	0	0	0.659	0	1	64	0.648	0	1	64
36	37–40	1792	2400	77.975	0	43	208.729	43	100	220	798.171	10	100	206
37	41–44	1658	2400	67.255	0	0	1.782	0	1	74	1.672	0	1	74
38	45–48	1820	2520	82.183	0	0	0.854	0	1	66	0.837	0	1	66
39	49–52	1482	2400	91.916	0	0	0.430	0	1	64	0.409	0	1	64

Table 5.14 Results for 6-week problems

Prob no.	Weeks	Total dem hours	Total sup hours	Time (s) start heuristic	Ini. LP soln	MPVarB Time (s) algo	Ini IP soln	Final IP soln	no. Nodes	no. Cols gen	SPVarB Time (s) algo	Ini IP soln	Final IP soln	no. Nodes	no. Cols gen
40	1–6	2,629	3,660	100.318	0	1.254	0	0	1	76	1.260	0	0	1	76
41	7–12	2,609	3,660	94.817	0	208.183	41	41	100	207	28.221	41	0	21	100
42	13–18	2,442	3,540	90.804	0	1.533	0	0	1	82	6.125	1,000	0	11	84
43	19–24	2,530	3,540	123.914	0	239.097	1,010	1,010	100	226	2.197	0	0	1	82
44	25–30	2,604	3,780	111.815	0	186.512	20	20	100	208	1.368	0	0	1	77
45	31–36	2,531	3,660	96.808	0	5.359	1,000	0	11	86	7.703	1,000	0	11	84
46	37–42	2,664	3,660	132.470	0	549.665	1,040	1,040	100	222	3,131.087	1,032	42	100	248
47	43–48	2,614	3,660	98.542	0	224.399	1,010	1,010	100	229	2,296.084	1,010	10	100	240

5.3.1 Two-Week Problems

In the first set of experiments, the year is broken down into 26 instances with 2-week planning horizons with the first instance starting on Monday 6th of January in 2005. On average, the number of regular demand periods $|T^{dem}| = 112$ (with a minimum of 95 and a maximum of 128) and the number of on-call periods $|\mathcal{L}| = 14$, so RMP has approximately 128 rows including 2 convexity constraints (48) for full-time and part-time physicians when 18 physicians are taken into account. In each instance, rosters are initialized with zero overtime hours. We do not use a rolling horizon approach because we are primarily interested in determining the difficulty in solving instances of fixed length. Including initial conditions does not change problem complexity nor did it affect runtimes in our preliminary testing.

Results are reported in Table 5.12. Column 1 gives the problem number and column 2 identifies the corresponding 2-week period during the year 2005. The next two columns give the total demand and the total supply. It is common in the service industry to have much more supply than demand since demand fluctuates over time (cf. Figs. 5.6–5.9) and labor agreements prohibit random assignments. In hospitals, it is critical that all demand is met which is the situation here.

Column 5 gives the time spent constructing the initial set of columns for RMP when the starting heuristic is employed. For a 2-week planning horizon, this required the solution of $n_{FT}^{phy} + n_{PT}^{phy} = 18$ integer programs with objective function (95). Although two sets of initial columns are needed for each of the 26 instances, depending on whether branching strategy MPVarB or SPVarB iss used, the time for the initialization heuristic is essentially the same in either case. Columns 6 and 7 provide the relaxed solutions to MP and the feasible solutions to MP found by solving it as an IP, respectively. In all 26 instances, the relaxed objective function value and the feasible objective function value are 0, indicating that optimality is achieved at the root node. In 10 of those instances, the LP solution is integral. The remaining 8 columns give the B&P statistics and solution values obtained with our two branching strategies. No branching is required for these instances. The four columns associated with either MPVarB or SPVarB give the time required by the B&P algorithm starting with the solution at the root node until termination, the final IP solution, the number of nodes explored, and the number of columns generated.

In all 26 instances in Table 5.12, either the RMP solution was integral or solving RMP as an IP after CGA has terminated at the root node has yielded the optimum. Because no branching is performed, the entries under MPVarB and SPVarB are identical with the exception of the runtimes, which are less than 1 s in all cases. The small differences are due to the fact that there is some randomness built into the procedure used by CPLEX to search for feasible solutions during B&B. With respect to the overall computational effort, solving the subproblems with objective function (95) in the initialization heuristic (cf. Sect. 4.3.7.1) is the dominant factor. Runtimes range from 2.44 to 27.54 s, averaging 9.72 s. When solving RMP at the root node, on average 48.08 columns are generated by CGA with a maximum of 53 for problem 2 and a minimum of 45 for problem 26.

5.3.2 Four-Week Problems

For the second set of experiments, the year is divided into 13 instances, each with a 4-week planning horizon. On average, RMP has a total of 255.69 constraints including 2 convexity constraints (48) for the two physician groups. Again, our goal is to compare MPVarB with SPVarB with respect to solution quality and runtimes. The results are summarized in Table 5.13.

The initialization heuristic averages 68.91 s for the 13 instances and produces the same set of columns for the MPVarB and SPVarB runs. Once again, the initialization component of the algorithm dominated the computational effort except for problem no. 36 which could not be solved to optimality with either MPVarB or SPVarB. The former could not improve on the initial LP solution of 43 within the 100 node limit, but SPVarB is able to obtain a value of 10. This corresponds to 1 h of planned overtime in four weeks which is acceptable for planners. Runtimes are approximately 209 and 800 s, and the number of columns generated is 220 and 206, respectively. In the two other cases where the initial IP solution is not optimal, both branching strategies are able to find an optimum. For problem no. 34, for example, 9 nodes are explored with MPVarB in 2 s and 6 nodes are explored with SPVarB in 2 s to arrive at a final solution with 0 cost.

5.3.3 Six-Week Problems

In the last set of experiments, we investigate 6-week problem instances. In these eight instances, RMP has an average of 370 rows. Table 5.14 summarizes the results for the eight problems solved. Rather than including a single column with heading "Initial IP soln", a pair of columns is included, one for each branching strategy. In these runs it is often the case that different feasible solutions are found at the root node because the columns in the RMP at termination are many times different.

As can be seen from the table, SPVarB is able to find the optimum in six of the eight problems whereas MPVarB has failed to converge in five instances. For problem nos. 40, 42, and 45 where both branching strategies find the optimum, MPVarB requires slightly less time. For problem nos. 46 and 47 where neither converges, SPVarB requires almost an order of magnitude more in time to reach the 100 node limit. This is a consequence of the relatively more complex constraints that are added to each subproblem when descendant nodes are created, especially on a right branch where each constraint includes a new binary variable. In the six instances where SPVarB does find the optimum, it does so quickly. By way of comparison, MPVarB is only able to improve on the initial IP solution for problem no. 45 although all runtimes where under 550 s.

5.3.4 General Observations

In general, we have found that the time spent on the initialization heuristic (cf. Sect. 4.3.7.1) is approximately the same for a given planning horizon, but in some instances, different feasible solutions are produced at the root node when RMP is solved as an IP. In all instances, though, many of the initial columns are in the final solutions, indicating that the effort to find good starting rosters can be worthwhile. Regarding the relative performance of the two branching schemes, SPVarB consumes much more time than MPVarB but is able to solve 45 of the 48 instances compared to 41 for MPVarB. The advantage of SPVarB is best evidenced when the planning horizon is 6 weeks, although both branching strategies show good performance when solving the 2- and 4-week instances. In most hospitals, planning is done in a rolling horizon framework, making short-term adjustments to account for absenteeism and personal requests. High quality solutions for up to 4 weeks at a time are generally what is needed.

5.4 Comparison of Both Algorithms

Recall, in Sect. 5.2.2 we have focused on the base case and on the investigation of changes in the parameter settings for the maximum shift length $\overline{T}_j^{shift}$, the time windows T_j^{win}, and the number of physicians that has been reduced to 13. The focus in this section is to solve the instances with our B&P algorithm. This is motivated by the fact that we could not solve all instances to optimality because mostly we could not improve on the lower bound in the root node of the original IP and the IP subproblems, respectively. Furthermore, remember that the decomposition is just an heuristic but our B&P algorithm is an exact solution methodology. We embody MPVarB when creating the search tree since we have found out in Sect. 5.3 that there is just a marginal difference between SPVarB and MPVarB when the planning horizon is short – in our case 2-week problems. Moreover, MPVarB reported faster runtimes than SPVarB. We apply the same settings as described in Sect. 5.3 except that we do not set a node limit on the search tree. Since we are applying an IP B&P algorithm it makes sense to use integer cost values (cf. Sect. 4.3.8.1). Hence, we scale the cost parameters to be used in a B&P framework and additionally we perform an appropriate scaling to be able to compare the results. We set $c^{out} = 1,000$, $c^{oc\_out} = 10,000$, $c^{over} = 10$, and $c^{paid} = 1$. Again, we use relative costs rather than absolute ones to facilitate the presentation. Table 5.15 shows the solution for the instances investigated. Column one indicates the problem instance. Column two to four characterizes the instance by maximum shift length, time window, and number of physicians. The next four columns repeat the solutions obtained in Sect. 5.2.2 for the different parameter constellations when the heuristic decomposition strategy has been employed. We report the best integer solution, the best lower bound, the optimality gap at termination, and the time spent to solve the

Table 5.15 Computational results different parameter settings

				Decomposition heuristic				MPVarB							
Prob no.	Max shift len.	Time win. len.	No. phy.	IP soln	Best lower bound	Gap (%)	Time (s) algo-rithm	IP soln	Best lower bound	Gap (%)	Time (s) start heuristic	Time (s) algo-rithm	Initial IP soln	No. nodes	No. cols gen
1	13	3	16	0	0	0	272	0	0	0	101	< 1	0	1	44
2	12	3	16	0	0	0	547	0	0	0	92	< 1	0	1	44
3	11	3	16	0	0	0	214	0	0	0	94	< 1	0	1	44
4	10	3	16	1,000	0	$-^a$	3,686	1,000	1,000	0	121	< 1	1,000	1	45
5	9	3	16	1,000	0	$-^a$	3,685	1,000	1,000	0	116	< 1	1,000	1	44
6	13	3	15	1,000	0	$-^a$	3,757	1,000	1,000	0	98	< 1	1,000	1	43
7	13	3	14	3,000	1,000	$-^a$	7,200	2,000	2,000	0	98	13	3,000	21	85
8	13	3	13	4,000	3,000	$-^a$	3,990	4,000	4,000	0	98	2,210	10,053	131	312
9	13	0	16	13,000	0	$-^a$	7,200	0	0	0	70	< 1	0	1	44
10	13	1	16	1,000	0	$-^a$	3,760	0	0	0	90	< 1	0	1	44
11	13	2	16	0	0	0	290	0	0	0	94	< 1	0	1	44

^{a}Optimum not found and no improvement in lower bound.

corresponding problem. Recall that the asterix indicates instances where we could not improve on the lower bound at least for one of the 2 weeks when applying the heuristic decomposition. The remaining eight columns detail the solution obtained by our B&P algorithm with MPVarB. The first three columns report the same information as for the decomposition heuristic. The next two columns show the time spend for the starting heuristic and the algorithm. The remaining columns report the initial feasible solution achieved at the root node, the generated shifts (columns), and the number of nodes explored.

General observations are: All instances are solved to optimality with the B&P algorithm in much less runtime, the lower bounds could be improved, the starting heuristic provides a good set of initial columns, and our procedure to find integer solutions performs well. In detail, nine of the eleven instances are solved in the root node where three instances have an integer solution when CGA has terminated with the node solution. Runtimes for the starting heuristic are relatively constant by approximately 100 s. This is because the second start of the procedure dominates the time spend (see Sect. 4.3.7.1). The B&P algorithm needs less than 1 s to prove optimality in that nine instances and in eight of the nine problems the initial set of columns is sufficient to proof the linear programming optimality at the root node. In problem eight one column is generated by CGA before the root solution is optimal.

For two problems, 7 and 8, branching is necessary to achieve an optimality gap of 0. For instance, consider problem 7 where an initial integer solution of 3,000 is found, 85 rosters are generated, and 21 nodes are explored before the B&P algorithm terminates after 111 s including the starting heuristic with an optimal solution of 2,000. Problem 8 is the hardest to solve. The B&P algorithm has needed 2,210 s to achieve optimality. 131 nodes are investigated and 312 columns are generated. The lower bound of 4,000 could be found after 46 iterations in the root node.

Figure 5.10 shows the progress of the LB. As can be seen, as soon as we are close to LP optimality in the root node CGA needs a lot of iterations, though just an marginal improvement is visible. This is the discussed tailing off effect column generation suffers. Additionally, in iteration two to five we have no improvement in the lower bound. Some researcher have a theory that column generation can be stopped if one cannot achieve a significant improvement over some iterations. The approach is justified with the occurrence of tailing off when the solution is near optimal.

Figure 5.11 shows the progress of the UB and the LB which is the horizontal line at hight 4,000 during the course of the B&P algorithm for problem instance 8. Almost every time we call our procedure to find feasible solutions (see Sect. 4.3.7.2) which is to solve MP as an IP, in our case at every tenth node, we find an improved UB. However, that approach consumes much runtime. At node 111 we achieve an almost optimal solution of 4,010 corresponding to 1 h of overtime in any roster. The optimality gap is almost closed, i.e. 0.25%. Upon that solution the algorithm needs about 500 s more to find the optimal solution at node 131 with cost of 4,000. It is doubtful if a optimal solution of 4,000 is preferable of a near optimal solution of 4,010 at cost of 500 s of runtime since we just consider planned overtime in our

5.4 Comparison of Both Algorithms

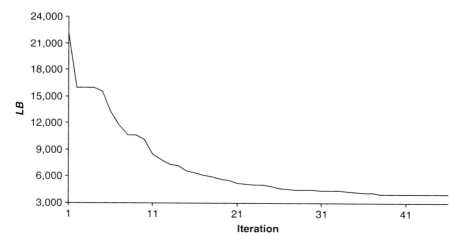

Fig. 5.10 Progress of LB in the root node for problem 8

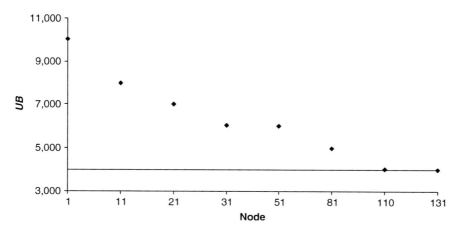

Fig. 5.11 Progress of UB in the course of the algorithm for problem 8

model. As the day of operation starts it might be likely that overtime is needed to cover unpredictable demand.

Finally, we conclude our investigation with the the result that our B&P algorithm with MPVarB performs very well in solving with up to 16 physicians and outperforms our decomposition heuristic with respect to solution quality and computational burden.

Chapter 6
Conclusions and Further Remarks

In the last chapter we summarize the thesis and draw conclusions in Sect. 6.1. In Sect. 6.2 we list some further research directions.

6.1 Summary and Conclusions

In this research we have considered the problem of flexible shift scheduling of physicians at a hospital. We have introduced a new modeling approach for this problem in Chap. 3 which allows shifts to be generated implicitly rather then using templates of, say, three 8-h shifts or two 12-h shifts to cover forecast demand. Furthermore, we have presented some enhancements like incorporation of on-call services or consideration of break assignments in the scheduling procedure among others, to solve more realistic problem instances. In Chap. 4 we have shown two solution methodologies. After presenting some preprocessing issues in Sect. 4.1 we first have introduced a heuristic decomposition strategy that divides the overall MIP into a sequence of weekly or multiple weekly subproblems. The reason why the decomposition is not exact is because decisions in previous weeks might effect solution quality in following weeks. However, the heuristic decomposition proofed to be effective when a weekly decomposition has been applied and it outperforms solving the overall MIP. Due to this drawback of the decomposition heuristic we have developed a new B&P algorithm that decomposes the underlying problem by physician or physician group to solve problems up to several weeks to optimality (see Sect. 4.3). There, we eliminate symmetry in the solution space by aggregating identical physicians in one subproblem in contrast to use one subproblem for each physician. In other words, we use general integer variables in the master problem formulation rather than binary decision variables which is more commonly known in the research literature. Two branching strategies have been applied. One concentrates on master problem variables and is easier to implement but leads to an unbalanced search tree. The other that deals with subproblem variables proofed to be more effective when longer planning horizons have been considered.

In Chap. 5 we perform three computational studies after stating the input data in Sect. 5.1. First, we have started investigating the decomposition heuristic in Sect. 4.2

J.O. Brunner, *Flexible Shift Planning in the Service Industry*, Lecture Notes in Economics and Mathematical Systems 640, DOI 10.1007/978-3-642-10517-3_6, © Springer-Verlag Berlin Heidelberg 2010

with respect to model features and parametric analysis. The results for the anesthesia department of the German university hospital MRI show that one can achieve high quality schedules in a reasonable amount of time when compared to current practice. The lines-of-work generated by our model allow full flexibility in terms of shift starting times and shifts lengths, and include the assignment of breaks, the use of planned overtime, and the need to follow the general labor agreement. In almost all cases, the schedules did not require planned overtime or the use of outside resources. In a second study we have demonstrated the ability of the proposed B&P algorithm to produce optimal schedules for up to 6 weeks at a time for both full-time and part-time physicians. A second contribution is centered on the adaptation and implementation of the Ryan and Foster branching scheme for set covering-type problems with integer variables and right-hand side values greater than one. We have referred to this scheme as SPVarB and have compared it with the more traditional MPVarB scheme that branches on MP variables. In all cases SPVarB provides equivalent or superior solutions but often with greater runtimes. For the 6-week instances, for example, runtimes averages 684.26 s compared to 177 s for MPVarB. Nevertheless, MPVarB provides acceptable performance and is competitive with SPVarB for short planning horizons. In a third study we have focused on resolving the parameter tests from study 1 with our B&P algorithm applying MPVarB. We have been able to solve all problems to optimality which was not possible with the decomposition strategy. Most of the problems investigated could be solved within 1 s since our starting heuristic provides a good set of initial rosters. From a computational point of view, we are able to generate tighter lower bounds at the root node which leads to an earlier termination of the algorithm with an optimal solution. Summarizing, the B&P algorithm needs much less runtime than the heuristic decomposition strategy and additionally provides optimal solutions. The results have led MRI to adopt our approach on a trial basis. For full implementation, it will be necessary to provide a graphical user interface that would permit the staff to enter their preferences and modify their schedules within the limits of local policies and the general labor agreement. A software tool that is connected to a database and uses the proposed solution methodologies is in development. One basic component for the success of the tool is a user interface that is easy to use by the physicians.

We close this section by stating the key contributions of the research which are:

- Formulation of an implicit model for flexible shift and break scheduling of physicians in hospitals that allows extraordinary flexibility in the scheduling process.
- Development and implementation of two efficient solution methods, namely a decomposition heuristic and an exact B&P algorithm.
- Successful testing on real world data provided by the anesthesia department at MRI.

6.2 Final Remarks and Further Research Directions

Because the environment in which physicians and other service sector workers operate is dynamic, this level of flexibility is critical in minimizing personnel costs. Flexibility also speaks to the desire of many supervisors to accommodate the requests and preferences of their employees, a basic component of job satisfaction. Nevertheless, the underlying problem is difficult to solve optimally for large instances, which is one of the reasons why there has been little work to date on physician scheduling. Further research should be aimed at balancing issues such as equal amount of overtime or on-call assignments amongst physicians and accounting for individual skill requirements and teaming objectives. As more constraints are added to the problem, the computational effort is likely to increase sharply. From a computational viewpoint, although some instances have been solved quickly, runtime variances are large for the different experiments. To improve overall efficiency, future research should take aim at developing more structured algorithms for solving the subproblems. For example, it may be possible to reformulate SP as network flow problem with side constraints and either adapt or develop a tailored dynamic programming algorithm to find solutions in the course of the B&P algorithm which increases the probability of pruning. Another research direction might concentrate on heuristics with the goal of finding high quality integer solutions. Hybrid schemes that employ metaheuristics to find initial solutions and then feasible solutions within B&B have proven to be the most effective way to solve large-scale scheduling and routing problems.

Anyway, allowing more flexibility as our work does in the scheduling procedures is crucial to tackle the volatile environment in which service organizations are operating. In the future it will be essential for survival that the budget is used more efficiently. One strategy is to employ more sophisticated scheduling procedures to reduce personnel cost which are the main cost block in the budget of a service organization – the focus of our research.

Appendix A
Appendix

A.1 Abbreviations, Notation, and Symbols

General abbreviations

AFT	Aggregated full-time physician
B&B	Branch and bound
B&P	Branch and price
CG	Column generation
CGA	Column generation algorithm
DMP	Dual master problem
FAZ	Frankfurter Allgemeine Zeitung
FT	Full-time physician
HNOP	Ear, nose and throat operating room
IP	Integer program
LB	Lower bound
LP	Linear program
MIP	Mixed integer program
MP	Master problem
MPVarB	Master problem variable branching
MRI	Muenchen Rechts der Isar (university hospital TUM)
PT	Part-time physician
REC	Roster elimination constraint
RMP	Restricted master problem
SP	Subproblem

SP_j	Subproblem for group j
SPOP	Sport operating room
SPVarB	Subproblem variable branching
UB	Upper bound
ZOP	Central operating room

General symbols

$\lfloor\ \ \rfloor$	Greatest integer less than or equal to argument
$\lceil\ \ \rceil$	Smallest integer greater than or equal to argument

Indices and sets

j	Index for physician groups
i	Index for physicians within a group
t, l	Index for periods
k	Index for schedules
w	Index for weeks
n	Index for nodes
u	Index for iterations in CGA
$\mathcal{J}$	Set of physician groups
$\mathcal{I}_j$	Set of physicians within a group j
$\mathcal{T}$	Set of periods in planning horizon
$\mathcal{T}_w$	Subset of periods within a week w with
	$\mathcal{T}_w = \{t \in \mathcal{T} : (w-1) \cdot T^{week} < t \le w \cdot T^{week}\} \subset \mathcal{T}$
$\mathcal{S}_w$	Set of first periods within any day in a week w with
	$\mathcal{S}_w = \{t \in \mathcal{T}_w : t \bmod T^{day} = 1\} \subset \mathcal{T}_w$
$\mathcal{T}^{dem}$	Set of periods where demand occurs with
	$\mathcal{T}^{dem} = \{t \in \mathcal{T} : d_t > 0\} \subset \mathcal{T}$
$\mathcal{T}^{earlystart}_{j,w}$	Set of earliest starting periods within
	the week w for physician group j
$\mathcal{W}$	Set of weeks in planning horizon
$\mathcal{L}$	Set of starting periods for an on-call service
$\mathcal{H}$	Set of holiday periods other than weekends with $\mathcal{H} \subset \mathcal{T}$
$\mathcal{V}_{j,i}$	Set of general vacation periods of physician i associated
	with group j where $\mathcal{V}_{j,i} \subset \mathcal{T}$

A.1 Abbreviations, Notation, and Symbols 105

$\mathcal{V}_{j,i}^{day}$	Set of general vacation days corresponding to set $\mathcal{V}_{j,i}$ of physician i associated with group j where $\mathcal{V}_{j,i}^{day} \subset \mathcal{L}$
$\mathcal{K}(j)$	Set of schedules for physician type $j \in \mathcal{J}$
$\mathcal{K}$	Set of all schedules with $\mathcal{K} = \bigcup_{j \in \mathcal{J}} \mathcal{K}(j)$
$\mathcal{S}^{shift}(j^*, k^*)$	Auxiliary set to perform MPVarB with $\mathcal{S}^{shift}(j^*, k^*) = \left\{ t \in \mathcal{T}^{dem} : X_t^{j^*, k^*} = 1 \right\}$
$\mathcal{S}^{break}(j^*, k^*)$	Auxiliary set to perform MPVarB with $\mathcal{S}^{break}(j^*, k^*) = \left\{ t \in \mathcal{T}^{dem} : Y_t^{j^*, k^*} = 1 \right\}$
$\mathcal{L}(j^*, k^*)$	Auxiliary set to perform MPVarB with $\mathcal{L}(j^*, k^*) = \left\{ t \in \mathcal{L} : Z_t^{j^*, k^*} = 1 \right\}$
$\hat{\mathcal{S}}^{shift}(j^*, k^*)$	Complement set to $\mathcal{S}^{shift}(j^*, k^*)$ with $\hat{\mathcal{S}}^{shift}(j^*, k^*) = \mathcal{T}^{dem} \setminus \mathcal{S}^{shift}(j^*, k^*)$
$\hat{\mathcal{S}}^{break}(j^*, k^*)$	Complement set to $\mathcal{S}^{break}(j^*, k^*)$ with $\hat{\mathcal{S}}^{break}(j^*, k^*) = \mathcal{T}^{dem} \setminus \mathcal{S}^{break}(j^*, k^*)$
$\hat{\mathcal{L}}(j^*, k^*)$	Complement set to $\mathcal{L}(j^*, k^*)$ with $\hat{\mathcal{L}}(j^*, k^*) = \mathcal{L} \setminus \mathcal{L}(j^*, k^*)$
$\mathcal{B}$	Set of all possible branching patterns when SPVarB is performed
$\mathcal{F}(j)$	Subset of branching patterns for group j with $\mathcal{F} \subset \mathcal{B}$
$\mathcal{N}(n)$	Set of nodes connecting node n with the root node
$\mathcal{P}(j)$	Set of physician type j with same working rules and different starting time windows

Parameters

d_t	Demand in period t
d_t^{oc}	Number of physicians required for each on-call service starting in period $t \in \mathcal{L}$
$c_{j,i}^{paid}$	Cost per hour of paid out time for physician i in group j
$c_{j,i}^{over}$	Cost per hour of overtime for physician i in group j
c_{jk}^{sched}	Cost associated with schedule (column) k for physician type j
c^{out}	Cost per hour for an outside physicians

c^{oc_out}	Cost per hour for an outside physicians
$\overline{c}$	Generic reduced cost
$\overline{o}_{j,i}$	Maximal allowed overtime for a physician i in group j in a week
$o_{j,i}^{prev}$	Initial overtime from the previous week for each $i \in \mathcal{I}_j$
$s_{j,i}^{prev}$	1, If physician $i \in \mathcal{I}_j$ associated with group $j \in \mathcal{J}$ is assigned an on-call service starting on the last day of the previous week, 0 otherwise
$\overline{T}_j^{shift}$	Maximum shift length for group j
$\underline{T}_j^{shift}$	Minimum shift length for group j
$\underline{T}_j^{rest}$	Minimum rest length after a shift ends for group j
$T_{j,w}^{startlength}$	Length of the starting time range
$r_{j,i}$	Regular working hours per week for physician i in group j according to general or individual agreement
r_i	Regular working hours per week according to the labor contract
T^{oc}	Length of an on-call service (periods)
T_{spill}^{oc}	Number of hours that an on-call service spills over to the following week
$\overline{n}_j^{oc}$	Maximum number of on-call allowed in a week
$\underline{T}_j^{pre}$	Minimum working periods before a break
$\overline{T}_j^{pre}$	Maximum working periods before a break
$\underline{T}_j^{post}$	Minimum working periods after a break
T^{week}	Number of periods within a week; $T^{week} = |\mathcal{T}|/|\mathcal{W}|$
T^{day}	Number of periods in a day; $T^{day} = T^{week}/7$
n_j^{phy}	Number of physician type j physicians that are available
$X_t^{j,k}$	1, If schedule k for type j covers regular demand in period $t \in \mathcal{T}^{dem}$, 0 otherwise
$Y_t^{j,k}$	1, If schedule k for type j has a break in period $t \in \mathcal{T}^{dem}$, 0 otherwise
$Z_t^{j,k}$	1, If schedule k for type j covers an on-call service starting in period $t \in \mathcal{L}$, 0 otherwise
δ_t^{dem}	Shadow prices of MP regular demand constraints for $t \in \mathcal{T}^{dem}$
δ_t^{oc}	Shadow prices of MP on-call demand constraints for $t \in \mathcal{L}$
δ_t^{dem}	Shadow prices of MP convexity constraints for $j \in \mathcal{J}$

A.1 Abbreviations, Notation, and Symbols

$\delta_{j,n}$ — Shadow prices for physician group j at node n when SPVarB

δ^{UB} — Dual variable when upper bound is imposed on one λ

δ^{LB} — Dual variable when lower bound is imposed on one λ

$\omega_{j,n}$ — Indicator parameter at node n when SPVarB is performed

α_{j^*,k^*} — Auxiliary variable to perform MPVarB

$$\alpha_{j^*,k^*} = \left|\mathcal{S}^{shift}\left(j^*,k^*\right)\right| + \left|\mathcal{S}^{rest}\left(j^*,k^*\right)\right| + \left|\mathcal{L}\left(j^*,k^*\right)\right|$$

z — Optimal objective function value for some optimization problem

Functions

$f_1(t)$ — Calculates the number of hours that are charged for any on-call service for each $t \in \mathcal{L}$ to regular working time per week

$f_2(t)$ — Maps each period $t \in \mathcal{T}$ into a period during a specific day, where $\mathcal{T} \longmapsto \{1, \dots, T^{day}\}$ with $t \mapsto f_2(t) = t \bmod T^{day}$

Binary decision variables

$x_{j,i,t}$ — 1, If physician i works in period t, 0 otherwise

$x_{j,t}$ — 1, If on duty in period t, 0 otherwise

$y^{shift}_{j,i,t}$ — 1, If physician i in group j begins a shift in period t, 0 otherwise

$y^{shift}_{j,t}$ — 1, If a shift begins in period t, 0 otherwise

$y^{rest}_{j,i,t}$ — 1, If rest period begins for physician i in group j in period t, 0 otherwise

$y^{rest}_{j,t}$ — 1, If rest period in period t, 0 otherwise

$y^{oc}_{j,i,t}$ — 1, If physician i in group j starts an on-call service in period $t \in \mathcal{L}$, 0 otherwise

$y^{oc}_{j,t}$ — 1, If an on-call service begins in period $t \in \mathcal{L}$, 0 otherwise

$y^{ind}_{j,i}$ — 1, If new column covers all elements in $\mathcal{F}(j)$ when SPVarB, 0 otherwise

General integer decision variables

$o_{j,i,w}$ — Amount of overtime for physician i in group j in week w

$o_{j,w}$ — Amount of overtime in week w

$u_{j,i,w}$	Amount of undertime for physician i in group j in week w
$u_{j,w}$	Amount of undertime in week w
$h_{j,i,w}$	Amount of paid out time for physician i in group j in week w
$h_{j,w}$	Amount of paid out time in week w
$e_{j,i,w}$	Earliest starting time for group j physician i in week w
$e_{j,w}$	Earliest starting time for a shift in week w
$l_{j,i,w}$	Latest starting time for group j physician i in week w
$l_{j,w}$	Latest starting time for a shift in week w
x_t^{out}	Number of outside physicians hours in period t
$x_t^{oc\_out}$	Number of outside physicians who carry out an on-call service starting in period $t \in \mathcal{L}$

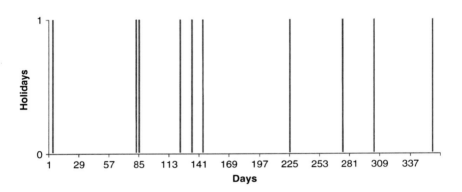

Fig. A.1 Holidays in 2005

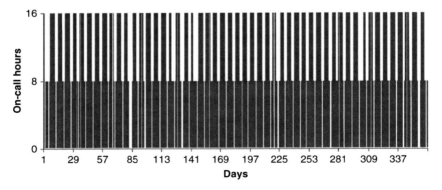

Fig. A.2 On-call hours counted to regular working hours for each on-call service in 2005

A.1 Abbreviations, Notation, and Symbols 109

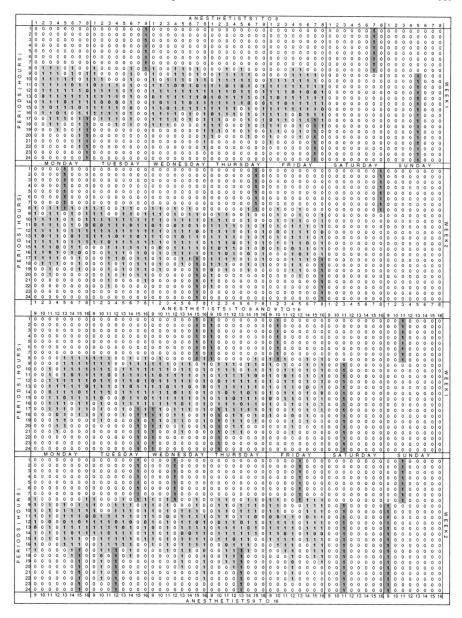

Fig. A.3 Duty roster produced by model for all anesthetists

Bibliography

1. Addou, I., Soumis, F.: Bechtold-jacobs generalized model for shift scheduling with extraordinary overlap. Ann. Oper. Res. **155**(1), 177–205 (2007)
2. Aiken, L.H., Clarke, S.P., Sloane, D.M., Sochalski, J., Silber, J.H.: Hospital nurse staffing and patient mortality, nurse burnout, and job dissatisfaction. J. Am. Med. Assoc. **288**, 1987–1993 (2002)
3. Al-Yakoob, S.M., Sherali, H.D.: A column generation approach for an employee scheduling problem with multiple shifts and work locations. J. Oper. Res. Society **59**, 34–43(10) (2008). DOI:10.1057/palgrave.jors.2602294. URL http://www.ingentaconnect.com/content/pal/01605682/2008/00000059/00000001/art00005
4. Alfares, H.K., Bailey, J.E.: Integrated project task and manpower scheduling. IIE Trans. **29**, 711–717 (1997). URL http://www.ingentaconnect.com/content/klu/iiet/1997/00000029/00000009/00173295
5. Arabeyre, J.P., Fearnley, J., Steiger, F.C., Teather, W.: The airline crew scheduling problem: A survey. Transp. Sci. **3**(2), 140–163 (1969). DOI:10.1287/trsc.3.2.140. URL http://transci.journal.informs.org/cgi/content/abstract/3/2/140
6. Azaiez, M.N., Sharif, S.S.A.: A 0-1 goal programming model for nurse scheduling. Comput. Oper. Res. **32**(3), 491–507 (2005). DOI:10.1016/S0305-0548(03)00249-1. URL http://www.sciencedirect.com/science/article/B6VC5-4BHV6G8-1/2/d5ed3b58b4f3cf93034decba5b49cbf9
7. Azmat, C.S., Hürlimann, T., Widmer, M.: Mixed integer programming to schedule a single-shift workforce under annualized hours. Ann. Oper. Res. **128**, 199–215 (2004). DOI:10.1023/B:ANOR.0000019105.54898.a4". URL http://www.ingentaconnect.com/content/klu/anor/2004/00000128/F0040001/05255906
8. Bailey, J.: Integrated days off and shift personnel scheduling. Comput. Ind. Eng. **9**(4), 395–404 (1985)
9. Bailey, J., Field, J.: Personnel scheduling with flexshift models. J. Oper. Manag. **5**(3), 327–338 (1985). DOI:10.1016/0272-6963(85)90017-8. URL http://www.sciencedirect.com/science/article/B6VB7-45K1NF6-1N/2/f3876a37d8684267033a2923a901fcbc
10. Bailyn, L., Collins, R., Song, Y.: Self-scheduling for hospital nurses: an attempt and its difficulties. J. Nurs. Manag. **15**, 72–77 (2007). DOI:10.1111/j.1365-2934.2006.00633.x. URL http://www.ingentaconnect.com/content/bsc/jnm/2007/00000015/00000001/art00010
11. Baker, K.R.: Scheduling a full-time workforce to meet cyclic staffing requirements. Manag. Sci. **20**(12), 1561–1568 (1974). DOI:10.1287/mnsc.20.12.1561. URL http://mansci.journal.informs.org/cgi/content/abstract/20/12/1561
12. Baker, K.R., Magazine, M.J.: Workforce scheduling with cyclic demands and day-off constraints. Manag. Sci. **24**(2), 161–167 (1977). DOI:10.1287/mnsc.24.2.161. URL http://mansci.journal.informs.org/cgi/content/abstract/24/2/161
13. Bard, J.F.: Staff scheduling in high volume service facilities with downgrading. IIE Trans. **36**(10), 985–997 (2004). URL http://search.ebscohost.com/login.aspx?direct=true&db=buh&AN=14361083&site=ehost-live

112 Bibliography

14. Bard, J.F., Binici, C., deSilva, A.H.: Staff scheduling at the united states postal service. Comput. Oper. Res. **30**(5), 745–771 (2003). DOI:10.1016/S0305-0548(02) 00048-5. URL http:// www.sciencedirect.com/science/article/B6VC5-45D8H6S-1/2/7dba9a 4c2f5b9514509fd17954d77e99

15. Bard, J.F., Morton, D.P., Wang, Y.M.: Workforce planning at USPS mail processing and distribution centers using stochastic optimization. Ann. Oper. Res. **155**(1), 51–78 (2007). URL http://dblp.uni-trier.de/db/journals/anor/anor155.html#BardMW07

16. Bard, J.F., Purnomo, H.W.: Preference scheduling for nurses using column generation. Eur. J. Oper. Res. **164**(2), 510–534 (2005). DOI:10.1016/j.ejor.2003.06.046. URL http://www.sciencedirect.com/science/article/B6VCT-4BYK0GN-1/2/fb33283ccbe144 36fe618e193c3fe64a

17. Bard, J.F., Purnomo, H.W.: Short-term nurse scheduling in response to daily fluctuations in supply and demand. Health Care Manag. Sci. **8**(4), 315–324 (2005). DOI:http://dx.doi.org/ 10.1007/s10729-005-4141-9. URL http://dx.doi.org/10.1007/s10729-005-4141-9

18. Bard, J.F., Purnomo, H.W.: Cyclic preference scheduling of nurses using a lagrangian-based heuristic. J. Scheduling **10**, 5–23 (2007). DOI:10.1007/s10951-006-0323-7. URL http:// www. ingentaconnect.com/content/klu/josh/2007/00000010/00000001/00000323

19. Bard, J.F., Wan, L.: Weekly scheduling in the service industry: An application to mail processing and distribution centers. IIE Trans. **37**, 379–396(18) (2005). DOI:10. 1080/07408170590885288. URL http://www.ingentaconnect.com/content/tandf/uiie/2005/ 00000037/ 00000005/art00001

20. Barnhart, C., Johnson, E.L., Nemhauser, G.L., Savelsbergh, M.W.P., Vance, P.H.: Branchand-price: Column generation for solving huge integer programs. Oper. Res. **46**, 316–329 (1998)

21. Bartholdi, J.J., Orlin, J.B., Ratliff, H.D.: Cyclic scheduling via integer programs with circular ones. Oper. Res. **28**(5), 1074–1085 (1980). DOI:10.1287/opre.28.5.1074. URL http:// or. journal.informs.org/cgi/content/abstract/28/5/1074

22. Beaulieu, H., Ferland, J.A., Gendron, B., Michelon, P.: A mathematical programming approach for scheduling physicians in the emergency room. Health Care Manag. Sci. **3**, 193–200 (2000). URL http://www.ingentaconnect.com/content/klu/hcms/2000/ 00000003/00000003/00327701

23. Beaumont, N.: Scheduling staff using mixed integer programming. Eur. J. Oper. Res. **98**(3), 473–484 (1997). DOI:10.1016/S0377-2217(97)00055-6. URL http:// www. sciencedirect.com/science/article/B6VCT-3SWY0P1-3/2/d53e5979293c4ea78e78ad635a28 90e7

24. Bechtold, S.E., Jacobs, L.E.: Implicit modeling of flexible break assignments in optimal shift scheduling. Manag. Sci. **36**(11), 1339–1351 (1990)

25. Bechtold, S.E., Jacobs, L.W.: Improvement of labour utilisation in shift scheduling for services with implicit optimal modelling. Int. J. Oper. Prod. Manage. **11**(2), 54–69 (1991). DOI:10.1108/EUM0000000001267

26. Bechtold, S.E., Jacobs, L.W.: The equivalence of general set-covering and implicit integer programming formulations for shift scheduling. Nav. Res. Logistics **2**(2), 233–249 (1996). DOI:10.1002/(SICI)1520-6750(199603)43:2⟨233::AID-NAV5⟩3.0.CO;2-B

27. Beliën, J.: Exact and heuristic methodologies for scheduling in hospitals: problems, formulations and algorithms. 4OR: Q. J. Oper. Res. **5**(2), 157–160 (2006). DOI:10.1007/ s10288-006-0006-4

28. Beliën, J., Demeulemeester, E.: Scheduling trainees at a hospital department using a branch-and-price approach. Eur. J. Oper. Res. **175**(1), 258–278 (2006). DOI:10.1016/j.ejor. 2005.04.028. URL http://www.sciencedirect.com/science/article/B6VCT-4GGWGCM-D/2/ 09fafee7ab899f6e41476aa39b51f68b

29. Beliën, J., Demeulemeester, E.: Building cyclic master surgery schedules with leveled resulting bed occupancy. Eur. J. Oper. Res. **176**(2), 1185–1204 (2007). DOI:10.1016/j. ejor.2005.06.063. URL http://www.sciencedirect.com/science/article/B6VCT-4HKMPSC-D/ 2/a4c0f441da18e2225a89382a6e22ed9e

Bibliography 113

30. Beliën, J., Demeulemeester, E.: A branch-and-price approach for integrating nurse and surgery scheduling. Eur. J. Oper. Res. **189**(3), 652–668 (2008). DOI:10.1016/j.ejor. 2006.10.060. URL http://www.sciencedirect.com/science/article/B6VCT-4MVDVHW-4/2/c4f6ee622310d2f1121f8906d48b15b3

31. Berrada, I., Ferland, J.A., Michelon, P.: A multi-objective approach to nurse scheduling with both hard and soft constraints. Socioecon. Plann. Sci. **30**(3), 183–193 (1996). URL http://www.sciencedirect.com/science/article/B6V6Y-3VWPXB5-3/2/be4e0e 945f0e6799c49fb8185e056ae4

32. Bester, M.J., Nieuwoudt, I., Vuuren, J.H.V.: Finding good nurse duty schedules: a case study. J. Scheduling **10**(6), 387–405 (2007). DOI:http://dx.doi.org/10.1007/s10951-007-0035-7

33. Bhushan, N., Rai, K.: Strategic decision making – applying the analytic hierarchy process. Springer (2004)

34. Blöchliger, I.: Modeling staff scheduling problems. a tutorial. Eur. J. Oper. Res. **158**(3), 533–542 (2004). DOI:10.1016/S0377-2217(03)00387-4. URL http://www.sciencedirect. com/science/article/B6VCT-49CT1N5-3/2/8a3410c21ee47564ab65a11946c2b78e

35. Bowman, E.H.: The schedule-sequencing problem. Oper. Res. **7**(5), 621–624 (1959). URL http://or.journal.informs.org/cgi/content/abstract/7/5/621

36. Bradley, D.J., Martin, J.B.: Continuous personnel scheduling algorithms: A literature review. J. Soc. Health Syst. **2**, 8–23 (1990)

37. Brusco, M.J.: Solving personnel tour scheduling problems using the dual all-integer cutting plane. IIE Trans. **30**, 835–844 (1998). URL http://www.ingentaconnect.com/content/klu/iiet/1998/00000030/00000009/00201527

38. Brusco, M.J., Jacobs, L.W.: Optimal models for meal-break and start-time flexibility in continuous tour scheduling. Manag. Sci. **46**(12), 1630–1641 (2000). DOI:http://dx.doi.org/10. 1287/mnsc.46.12.1630.12074

39. Burke, E., Cowling, P., De Causmaecker, P., Vanden Berghe, G.: A memetic approach to the nurse rostering problem. Appl. Intell. **15**, 199–214 (2001). URL http://www.ingentaconnect. com/content/klu/apin/2001/00000015/00000003/00354286

40. Burke, E., De Causmaecker, P., Petrovic, S., Vanden Berghe, G.: Metaheuristics for handling time interval coverage constraints in nurse scheduling. Appl. Artif. Intell. Int. J. **20**(9), 743 766 (2006)

41. Burke, E.K., De Causmaecker, P., Vanden Berghe, G., Van Landeghem, H.: The state of the art in nurse rostering. J. Scheduling **7**, 441–499 (2004). DOI:10.1023/B:JOSH.0000046076. 75950.0b

42. Burns, R., Narasimhan, R.: Multiple shift scheduling of workforce on four-day workweeks. J. Oper. Res. Soc. **50**, 979–981 (1999). URL http://www.ingentaconnect.com/content/pal/01605682/1999/00000050/00000009/2600726

43. Burns, R.N., Carter, M.W.: Work force size and single shift schedules with variable demands. Manag. Sci. **31**(5), 599–607 (1985). URL http://search.ebscohost.com/login.aspx? direct=true&db=buh&AN=7355791&site=ehost-live

44. Cai, X., Li, K.N.: A genetic algorithm for scheduling staff of mixed skills under multi-criteria. Eur. J. Oper. Res. **125**(2), 359–369 (2000). URL http://ideas.repec.org/a/eee/ejores/v125y2000i2p359-369.html

45. Canon, C.: Personnel scheduling in the call center industry. 4OR: Q. J. Oper. Res. **5**(1), 89–92 (2007). DOI:10.1007/s10288-006-0008-2. URL http://www.springerlink.com/content/ch3mtm5kn4435h61

46. Carter, M.W., Lapierre, S.D.: Scheduling emergency room physicians". Health Care Manag. Sci. **4**, 347–360 (December 2001). URL http://www.ingentaconnect.com/content/klu/hcms/2001/00000004/00000004/00357207

47. Çezik, T., Günlük, O., Luss, H.: An integer programming model for the weekly tour scheduling problem. SO: Nav. Res. Logistics **48**(7), 607–624 (2001). DOI:10.1002/nav.1037. URL http://dx.doi.org/10.1002/nav.1037

48. Ceselli, A., Righini, G.: A branch-and-price algorithm for the multilevel generalized assignment problem. Oper. Res. **54**(6), 1172–1184 (2006). DOI:http://dx.doi.org/10.1287/opre. 1060.0323

49. Cheang, B., Li, H., Lim, A., Rodrigues, B.: Nurse rostering problems–a bibliographic survey. Eur. J. Oper. Res. **151**(3), 447–460 (2003). DOI:10.1016/S0377-2217(03)00021-3. URL http://www.sciencedirect.com/science/article/B6VCT-487N2B9-5/2/bb155 37b34313a93c74ea9d531c7a024
50. Chiaramonte, M.V., Chiaramonte, L.M.: An agent-based nurse rostering system under minimal staffing conditions. Int. J. Prod. Econ. **114**(2), 697–713 (2008). URL http://ideas.repec.org/a/eee/proeco/v114y2008i2p697-713.html
51. Chvátal, V.: Linear programming. W. H. Freeman, New York (1983)
52. Cohn, A., Root, S., Esses, J., Kymissis, C., Westmoreland, N.: Using mathematical programming to schedule medical residents. Technical Report (2006)
53. Dantzig, G.B., Wolfe, P.: Decomposition principle for linear programs. Oper. Res. **8**(1), 101–111 (1960). URL http://www.jstor.org/stable/167547
54. De Causmaecker, P., Demeester, P., Vanden Berghe, G., Verbeke, B. (eds.): Analysis of real-world personnel scheduling problems. Proceedings of the 5th international conference on practice and theory of automated timetabling, Pittsburgh (2004)
55. Desaulniers, G.: Managing large fixed costs in vehicle routing and crew scheduling problems solved by column generation. Comput. Oper. Res. **34**(4), 1221–1239 (2007). DOI:10.1016/j.cor.2005.07.002. URL http://www.sciencedirect.com/science/article/B6VC5-4GYH81X-1/2/62008f13073d4819b6409c62df7ffc43
56. Desaulniers, G., Desrosiers, J., Solomon, M.M.: Accelerating strategies in column generation methods for vehicle routing and crew scheduling problems. Groupe d'etudes et de recherche en analyse des decisions, Montréal (1999)
57. Desaulniers, G., Desrosiers, J., Solomon, M.M.: Column generation. Springer, New York (2005)
58. Desrosiers, J.: Time constrained routing and scheduling. Groupe d'etudes et de recherche en analyse des decisions, Montréal (1994)
59. Dowling, D., Krishnamoorthy, M., Mackenzie, H., Sier, D.: Staff rostering at a large international airport. Ann. Oper. Res. **72**(0), 125–147 (1997). DOI:10.1023/A:1018992120116. URL http://www.springerlink.com/content/n426qw3202375435
60. Dowsland, K.A.: Nurse scheduling with tabu search and strategic oscillation. Eur. J. Oper. Res. **106**(2-3), 393–407 (1998). DOI:10.1016/S0377-2217(97)00281-6. URL http://www.sciencedirect.com/science/article/B6VCT-3V7B1NM-C/2/d51cadd1561179877d 1b9d65db68109e
61. Dumas, Y., Desrosiers, J., Soumis, F.: The pickup and delivery problem with time windows. Eur. J. Oper. Res. **54**(1), 7–22 (1991). URL http://www.sciencedirect.com/science/article/B6VCT-48VW7F7-80/2/c885cb26c85dec2e6a81db13d8c4336c
62. Easton, F.F., Rossin, D.F.: Sufficient working subsets for the tour scheduling problem. Manag. Sci. **37**(11), 1441–1451 (1991)
63. Easton, F.F., Rossin, D.F.: Overtime schedules for full-time service workers. Omega **25**(3), 285–299 (1997). URL http://www.sciencedirect.com/science/article/B6VC4-3SWY81Y-3/2/7941dc9d7bb00cfeb4a32673a2864abf
64. Ernst, A.T., Jiang, H., Krishnamoorthy, M., Owens, B., Sier, D.: An annotated bibliography of personnel scheduling and rostering. Ann. Oper. Res. **127**(1), 21–144 (2004). DOI:10.1023/B:ANOR.0000019087.46656.e2. URL http://www.springerlink.com/content/j15j158945723185
65. Ernst, A.T., Jiang, H., Krishnamoorthy, M., Sier, D.: Staff scheduling and rostering: A review of applications, methods and models. Eur. J. Oper. Res. **153**(1), 3–27 (2004). DOI:10.1016/S0377-2217(03)00095-X. URL http://www.sciencedirect.com/science/article/B6VCT-48BKYWR-4/2/4b065270117e5421bb1299161a78803d. Timetabling and Rostering
66. Eveborn, P., Ronnqvist, M.: Scheduler - a system for staff planning. Ann. Oper. Res. **128**, 21–45 (2004). DOI:10.1023/B:ANOR.0000019097.93634.07. URL http://www.ingentaconnect.com/content/klu/anor/2004/00000128/F0040001/05255895
67. Farley, A.A.: A note on bounding a class of linear programming problems, including cutting stock problems. Oper. Res. **38**(5), 922–923 (1990). URL http://search.ebscohost.com/login.aspx?direct=true&db=buh&AN=4495224&site=ehost-live

Bibliography 115

68. Felici, G., Gentile, C.: A polyhedral approach for the staff rostering problem. Manag. Sci. **50**(3), 381–393 (2004). DOI:10.1287/mnsc.1030.0142. URL http://mansci.journal.informs. org/cgi/content/abstract/50/3/381
69. Ford, L.R., Fulkerson, D.R.: A suggested computation for maximal multi-commodity network flows. Manag. Sci. **50**(12), 1778–1780 (2004). DOI:http://dx.doi.org/10.1287/mnsc. 1040.0269
70. Franz, L.S., Miller, J.L.: Scheduling medical residents to rotations: Solving the large-scale multiperiod staff assignment problem. Oper. Res. **41**(2), 269–279 (1993). DOI:10.1287/opre. 41.2.269. URL http://or.journal.informs.org/cgi/content/abstract/41/2/269
71. Gamache, M., Soumis, F., Marquis, G.: A column generation approach for large-scale aircrew rostering problems. Oper. Res. **47**(2), 247–263 (1999). DOI:10.1287/opre.47.2.247. URL http://or.journal.informs.org/cgi/content/abstract/47/2/247
72. Geoffrion, A.M.: Lagrangean relaxation for integer programming. Math. Program. Stud. **2**, 82–114 (1974)
73. Gilmore, P.C., Gomory, R.E.: A linear programming approach to the cutting-stock problem. Oper. Res. **9**(6), 849–859 (1961). DOI:http://dx.doi.org/10.2307/167051. URL http://dx.doi. org/10.2307/167051
74. Gilmore, P.C., Gomory, R.E.: A linear programming approach to the cutting-stock problem – Part II. Oper. Res. **11**(6), 863–888 (1963). DOI:10.1287/opre.11.6.863. URL http://or.journal. informs.org/cgi/content/abstract/11/6/863
75. Göbel, M., Franck, M., Friesdorf, W.: Work duration and work scheduling for hospital doctors. Qual. Work Prod. Enterp. Future **31**, 667–708 (2003)
76. Hans, E.W.: Resource loading by branch-and-price techniques. Ph.D. thesis, University of Twente, Enschede (2001). URL http://doc.utwente.nl/36552/
77. Hochbaum, D.S., Levin, A.: Cyclical scheduling and multi-shift scheduling: Complexity and approximation algorithms. Discrete Optim. **3**(4), 327–340 (2006). DOI:10.1016/j.disopt. 2006.02.002. URL http://www.sciencedirect.com/science/article/B7GWV-4KD5BYK-3/2/ f8aa099606b1f6161092e3ca6c9e05c4
78. Hoffmann, K.L., Padberg, M.: Solving airline crew scheduling problems by branch-and-cut. Manag. Sci. **39**(6), 657–682 (1993)
79. Ingolfsson, A., Haque, M.A., Umnikov, A.: Accounting for time-varying queueing effects in workforce scheduling. Eur. J. Oper. Res. **139**(3), 585–597 (2002). URL http://ideas.repec.org/ a/eee/ejores/v139y2002i3p585-597.html
80. Isken, M.W.: An implicit tour scheduling model with applications in healthcare. Ann. Oper. Res. **128**(1), 91–109 (2004). DOI:10.1023/B:ANOR.0000019100.08333.a7. URL http:// www.springerlink.com/content/w052q417876765mm
81. Jaumard, B., Semet, F., Vovor, T.: A generalized linear programming model for nurse scheduling. Eur. J. Oper. Res. **107**(1), 1–18 (1998). URL http://ideas.repec.org/a/eee/ejores/ v107y1998i1p1-18.html
82. Kellogg, D.L., Walczak, S.: Nurse scheduling: From academia to implementation or not? Interfaces **37**(4), 355–369 (2007). URL http://interfaces.journal.informs.org/cgi/content/ abstract/37/4/355
83. Kohl, N., Karisch, S.E.: Airline crew rostering: Problem types, modeling, and optimization. Ann. Oper. Res. **127**, 223–257 (2004). DOI:10.1023/B:ANOR.0000019091.54417.ca. URL http://www.ingentaconnect.com/content/klu/anor/2004/00000127/F0040001/05255887
84. Lasdon, L.S.: Optimization theory for large systems. Macmillan, London (1970)
85. Lavoie, S., Minoux, M., Odier, E.: A new approach for crew pairing problems by column generation with an application to air transportation. Eur. J. Oper. Res. **35**(1), 45–58 (1988). URL http://ideas.repec.org/a/eee/ejores/v35y1988i1p45-58.html
86. Lübbecke, M.E., Desrosiers, J.: Selected topics in column generation. Oper. Res. **53**(6), 1007–1023 (2005). DOI:http://dx.doi.org/10.1287/opre.1050.0234
87. Mandelbaum, A.: Call centers (centres): Research bibliography with abstracts. **Version 3** (2006)

88. Marburger Bund und Tarifgemeinschaft deutscher Länder: Der arztspezifische Tarifvertrag für die Universitätsärzte (2007). URL http://www.marburger-bund.de/marburgerbund/bundesverband/unsere_themen/tarifpolitik/tdl/tdl-028.php
89. Mc Manus, M.I.: Optimum use of overtime in post offices. Comput. Oper. Res. **4**(4), 271–278 (1977)
90. Mehrotra, A., Murphy, K.E., Trick, M.A.: Optimal shift scheduling: A branch-and-price approach. Nav. Res. Logistics **47**(3), 185–200 (2000). DOI:10.1002/(SICI)1520-6750(200004)47:3⟨185::AID-NAV1⟩3.0.CO;2-7. URL http://dx.doi.org/10.1002/(SICI)1520-6750(200004)47:3⟨185::AID-NAV1⟩3.0.CO;2-7
91. Mihm, A.: Krankenhäuser schlagen Alarm. Frankfurter Allgemeine Zeitung (2007). URL http://berufundchance.fazjob.net/s/RubC43EEA6BF57E4A09925C1D802785495A/Doc~EB9067149F560459D998C1E8F7DE953E6~ATpl~Ecommon~Scontent.html
92. Millar, H.H., Kiragu, M.: Cyclic and non-cyclic scheduling of 12 h shift nurses by network programming. Eur. J. Oper. Res. **104**(3), 582–592 (1998). DOI:10.1016/S0377-2217(97)00006-4. URL http://www.sciencedirect.com/science/article/B6VCT-3SX6GTB-1N/2/c1f36feec87a1b623cc3b432c8c216a8
93. Moondra, S.L.: An LP model for work force scheduling for banks. J. Bank Res. **7**, 299–301 (1976)
94. Mundschenk, M., Drexl, A.: Workforce planning in the printing industry. Int. J. Prod. Res. **45**(20), 4849–4872 (2007)
95. Ni, H., Abeledo, H.: A branch-and-price approach for large-scale employee tour scheduling problems. Ann. Oper. Res. **155**(1), 167–176 (2000). DOI:10.1007/s10479-007-0212-2. URL Link-http://www.springerlink.com/content/k7w42v8028305j40
96. O'Neil, E., Seago, J.A.: Meeting the challenge of nursing and the nation's health. J. Am. Med. Assoc. **288**(16), 2040–2041 (2002). URL http://jama.ama-assn.org
97. Ovchinnikov, A., Milner, J.: Spreadsheet model helps to assign medical residents at the university of vermont's college of medicine. Interfaces **38**(4), 311–323 (2008). DOI:10.1287/inte.1070.0337
98. Pritsker, A.A.B., Watters, L.J., Wolfe, P.M.: Multiproject scheduling with limited resources: A zero-one programming approach. Manag. Sci. **16**(1), 93–108 (1969). URL http://search.ebscohost.com/login.aspx?direct=true&db=buh&AN=7107181&site=ehost-live
99. Purnomo, H.W., Bard, J.F.: Cyclic preference scheduling for nurses using branch and price. Nav. Res. Logistics **54**(2), 200–220 (2007)
100. Rekik, M., Cordeau, J.F., Soumis, F.: Using benders decomposition to implicitly model tour scheduling. Ann. Oper. Res. **128**, 111–133(23) (2004). DOI:10.1023/B:ANOR.0000019101.29692.2c. URL http://www.ingentaconnect.com/content/klu/anor/2004/00000128/F0040001/05255899
101. Rousseau, L.M., Pesant, G., Gendreau, M.: A general approach to the physician rostering problem. Ann. Oper. Res. **115**, 193–205 (2002). DOI:10.1023/A:1021153305410. URL http://www.springerlink.com/content/pr4547q2p01212g4
102. Ryan, D.M., Foster, B.A.: An integer programming approach to scheduling. In: Wern, A. (ed.) Computer scheduling of public transport urban passenger vehicle and crew scheduling, North-Holland, Amsterdam (1981)
103. Sarin, S.C., Aggarwal, S.: Modeling and algorithmic development of a staff scheduling problem. Eur. J. Oper. Res. **128**(3), 558–569 (2001). DOI:10.1016/S0377-2217(99)00421-X. URL http://www.sciencedirect.com/science/article/B6VCT-41TN4MN-7/2/279c8210d528a551c2f911af33f63b5b
104. Savelsbergh, M.: A branch-and-price algorithm for the generalized assignment problem. Oper. Res. **45**(6), 831–841 (1997)
105. Schniederjans, M.J.: Goal programming: Methodology and applications. Kluwer, Boston (1995)
106. Sherali, H.D., Ramahi, M.H., Saifee, Q.J.: Hospital resident scheduling problem. Prod. Plann. Control **13**(2), 220–233 (2002)
107. Sitompul, D., Randhawa, S.U.: Nurse scheduling models - A state of the art review. J. Soc. Health Syst. **2**(1), 62–72 (1990)

Bibliography

108. Sol, M.: Column generation techniques for pickup and delivery problems. Ph.D. thesis, Technische Universität Eindhoven (1994)
109. Soumis, F.: Decomposition and column generation. Groupe d'etudes et de recherche en analyse des decisions, Montréal (1997)
110. Thompson, G.M.: Improved implicit optimal modeling of the labor shift scheduling problem. Manag. Sci. **41**(4), 595–607 (1995)
111. Thungjaroenkul, P., Cummings, G.G., Embleton, A.: The impact of nurse staffing on hospital costs and patient length of stay: A systematic review. Nurs. Econ.\$ **25**(5), 255–265 (2007). URL http://www.find-health-articles.com/rec_pub_18080621-the-impact-nurse-staffing-hospital-costs-patient-length-stay.htm
112. Tien, J.M., Kamiyama, A.: On manpower scheduling algorithms. Soc. Ind. Appl. Math. Rev. **24**(3), 275–287 (1982). DOI:10.1137/1024063. URL http://link.aip.org/link/?SIR/24/275/1
113. Topaloglu, S.: A multi-objective programming model for scheduling emergency medicine residents. Comput. Ind. Eng. **51**(3), 375–388 (2006). DOI:10.1016/j.cie.2006.08.003. URL http://www.sciencedirect.com/science/article/B6V27-4KW5W8J-5/2/8d28377700507de24b1dc125b7861979. Special Issue on Selected Papers from the 34th. International Conference on Computers and Industrial Engineering (ICC&IE)
114. Topaloglu, S.: A shift scheduling model for employees with different seniority levels and an application in healthcare. Eur. J. Oper. Res. **In Press** (2008). DOI:10.1016/j.ejor.2008.10.032. URL http://www.sciencedirect.com/science/article/B6VCT-4TWV6MD-1/2/0876c76a0f88b8ff61016bdd2c9ceb09
115. Vanderbeck, F.: Decomposition and column generation for integer programs. Ph.D. thesis, Universite Catholique de Louvain (1994)
116. Vanderbeck, F.: On dantzig-wolfe decomposition in integer programming and ways to perform branching in a branch-and-price algorithm. Oper. Res. **48**(1), 111–128 (2000). DOI: http://dx.doi.org/10.1287/opre.48.1.111.12453
117. Vanderbeck, F., Wolsey, L.A.: An exact algorithm for IP column generation. Oper. Res. Letters **19**(4), 151–159 (1996). DOI:10.1016/0167-6377(96)00033-8. URL http://www.sciencedirect.com/science/article/B6V8M-3VWTCWR-1/2/41af45df4032495c596d0c20855ac344
118. Wan, L., Bard, J.F.: Weekly staff scheduling with workstation group restrictions. J. Oper. Res. Soc. **58**(8), 1030–1046 (2007)
119. White, C.A., White, G.M.: Scheduling doctors for clinical training unit rounds using tabu optimization. Pract. Theory Autom. Timetabling **IV**, 120–128 (2003). URL http://www.springerlink.com/content/khr1pxn918eqmxuk
120. Wilhelm, W.E.: A technical review of column generation in integer programming. Optim. Eng. **2**, 159–200 (2001). DOI:10.1023/A:1013141227104. URL http://www.springerlink.com/content/x820pp778h158l01
121. Williams, H.P.: Model building in mathematical programming, 4th edn. Wiley, Chichester (2001)
122. Winston, W.L.: Operations research applications and algorithms, 4th edn. Thomson Brooks/Cole, Belmont (2004)
123. Wolsey, L.A.: Integer programming. Wiley, New York (1998)

Breinigsville, PA USA
25 March 2010

234871BV00004B/62/P